HAZARDOUS MATERIALS CHARACTERIZATION

HAZARDOUS MATERIALS CHARACTERIZATION
Evaluation Methods, Procedures, and Considerations

Donald A. Shafer

WILEY-INTERSCIENCE

A JOHN WILEY & SONS, INC., PUBLICATION

For general information on our other products and services or for technical support, please contact our
Customer Care Department within the United States at (800) 762-2974, outside the United States at
(317) 572-3993 or fax (317) 572-4002.

Wiley also publishes its books in a variety of electronic formats. Some content that appears in print may
not be available in electronic formats. For more information about Wiley products, visit our web site at
www.wiley.com.

Library of Congress Cataloging-in-Publication data is available.

ISBN-13 978-0-471-46257-6
ISBN-10 0-471-46257-8

Printed in the United States of America.

10 9 8 7 6 5 4 3 2 1

To Karen, who makes our house a home.
Eternal Companion, Mother, and Friend

ACKNOWLEDGMENTS

Special thanks to the kind folks at the Hazardous Materials Training and Research Institute (HMTRI) at Kirkwood Community College, in Cedar Rapids, Iowa for the valuable training and professional trainer support that I have received from them and for the government training support programs that they administer to me and to hundreds of my colleagues throughout the nation. Thanks also to my many professional associates and clients that have graciously allowed me to work with them throughout my career.

ABOUT THE AUTHOR

Don Shafer has been a hazardous material professional for over 30 years. His career spans many years of industrial chemical process engineering experience, emergency planning and response, working as a "mad scientist," college instructor, consultant, and training the troops in the trenches. He is a degreed Microbiologist and Chemist. He is a Certified Environmental, Health and Safety Trainer and Registered Environmental Professional. For many years he served on the Board of Directors and was elected to President of the National Environmental, Health and Safety Training Association. He is a patriot, deeply loves God and America. He is a veteran of the Vietnam War.

Most importantly, he is a dedicated husband, father, and grandfather.

CONTENTS

1 REGULATORY CONTROL OF DANGEROUS SUBSTANCES   **1**

Regulatory Control–the Code of Federal Regulations, 3

Environmental Protection Agency (EPA), 4

    I. EPA: Clean Water Act, 5

    II. Clean Air Act (CAA), 5

       National Ambient Air Quality Standards (NAAQS),

       National Emission Standards For Hazardous Air Pollutants, 6

       Major Provisions of the Clean Air Act, 6

    III. Safe Drinking Water Act (SDWA), 7

       Public Water Management, 7

    IV. Hazardous Waste Management Resource Conservation And Recovery Act (RCRA), 7

       A. Major Elements Of RCRA Regulations, 7

       B. Key RCRA Definitions, 8

       C. Hazardous Waste Identification, 8

       D. Reportable Quantities (RQ), 9

       E. Characterization of Hazardous Wastes, 9

       F. Hazardous Waste Management Requirement, 10

          Regulatory Agency Notificationv10

          Hazardous Waste Permits (40 CFR 270), 10

          Hazardous Waste Tracking, 10

          Hazardous Waste Reporting (40 CFR 262-265), 10

          Record Keeping (262-265), 10

V. Superfund Acts, 10

CERCLA, 11

Superfund Amendment And Reauthorization Act (SARA), 11

Title I, 11

Title II, 12

VI. Emergency Planning and Community Right to Know (EPCRA), 12

Emergency Planning and Notification, 12

Reporting and Notification, 13

Comprehensive Emergency Response Plans, 13

VII. Pollution Prevention Act, 13

Pollution Prevention Ethic, 13

VIII. Occupational Safety and Health Act (OSH Act), 14

Safety and Health Standards, 14

Who Is NIOSH?, 14

OSHA Requirements for Employers, 14

1. Examples of Considerations for Evaluating Safe Work Places, 14

2. OSHA Enforcement, 16

3. OSHA Hazard Communication Program, 16

Key Elements of HAZCOMM, 16

4. Informing Workers About Workplace Hazards, 17

2 **PRINCIPLES OF SAFETY**   19

Safe Work Practices, 19

OSHA Self Evaluation, 19

Regulatory Programs Pertaining to Hazardous Materials Work Sites, 20

Considerations for Safe Work Practices, 21

Communicating Hazards Information, 22

Employee Right to Know, 22

Community Right to Know, 22

Walking and Working Surfaces, 23

Overhead and Underground Utilities, 23

Electrical Safety, 23

Tools and Heavy Equipment, 24

Protection from Hazardous Energy, 24

Lifting and Repetitive Motions, 25

Human Blood and Body Fluids, 26

Biohazards, 27

Hazardous Materials Spills and Releases, 27

Recognizing Hazards, 29

Site Specific Health and Safety Plans (HASP), 29

Recognizing Hazards Related to Dangerous Materials, 30

   Chemical Hazard Communication Program, 30

Community Right to Know, 32

   What Are Emergency Planning and Community Right to Know Issues, 32

Recognition of Chemical Hazards in Shipping and Handling, 34

Hazard Recognition Summary, 34

Hazard Recognition Discussion, 38

   Physical and Site Hazards, 38

   Personal Health Status, 38

   Improper Personal Protection, 38

   Lack of Oxygen, 39

   Too Much Oxygen, 39

   Space Configuration, 39

   Site Configuration and Layout, 39

   Environmental Hazards, 39

   Equipment Operation, 40

   Improperly Maintained Worksites, 40

   Defective Materials and Structural Integrity, 40

   Walking and Working Surfaces—Slip, Trip, and Fall Prevention, 41

   Overhead Activities, 41

   Improper Lifting and Ergonomics Practices, 41

   Electrical Hazards, 41

   Ventilation and Noise Hazards, 41

   Pneumatic and Fluid Pressure Hazards, 42

   Temperature Extremes, 42

   Acts of Terrorism and Sabotage, 42

Noncompliance with Regulatory Standards, 42

Improper Management Practices, Improper Work Practices, and Improper Standard Operating Procedures, 43

Lack of Discipline and Training, 43

About Training, 43

Personal Protection, 45

PPE Hazard Assessment, 45

PPE Training, 45

Head Protection, 45

Selection of Head Protection, 46

Eye and Face Protection, 46

Hearing Protection, 47

Chemical Protective Clothing (CPC), 47

PPE Selection, 47

Conditions Requiring Special Personal Protective Equipment, 48

Levels of Protection, 49

Flash Protective Suits, 50

Respiratory Protection, 51

Respiratory Protection Program, 51

Kinds of Respirators, 51

Standard Operating Procedures, 55

Safety and Health Management Program, 55

A. Purpose and Scope, 55

B. Applicability, 55

C. Reference, 55

D. General, 55

E. Management Leadership and Employee Participation, 56

F. Employee Participation, 56

G. Hazard Identification and Assessment, 57

H. Hazard Prevention and Control, 57

I. Information and Training, 58

J. Evaluation of Program Effectiveness, 58

K. Multi-Employer Workplace or Worksite, 59

L. Responsibilities, 59

3   DANGEROUS SUBSTANCES   **61**

Properties of Dangerous Substances, 61

Material Forms, 62

Corrosive Hazards, 63

  What About pH?, 64

  Corrosive Hazards—Volatility and Reactivity, 65

  Reducing Hazardous Properties of Corrosive Substances, 65

Electronic Charges and Energies of Attraction - The Glue That Holds Everything Together, 66

  Acids, 66

    About Acid pH, 66

  Alkali, 67

    About Alkali pH, 67

What about Neutralization?, 67

  Neutralize Using a Weak Opposite, 68

    Neutralizing Hydrochloric Acid (H+ Cl -) (pH ranges less than 2), 68

    Neutralizing Caustic Soda (NaOH) (pH greater than 12.5), 69

Ignitable Hazards, 69

  Percent to PPM, 70

  Vapor Pressure, 70

    Vapor Density and Inadequate Ventilation, 71

Toxic and Health Hazards, 71

Human Health Considerations, 72

  Critical Competencies, 73

Health Conditions Related to Dangerous Substance Exposure, 73

  Irritation, 74

  Synergism, 74

  Sensitization, 74

  Asphyxiation, 74

  Respiratory Paralysis, 75

  Systemic Poisoning, 75

Mutagenesis, 75

Teratogenicity, 76

Carcinogenesis, 76

Systemic Toxicity, 77

Organ Systems, 77

Toxic Substances, 78

Typical Pesticides, 78

Insecticides, 78

Herbicides, 79

Fungicides, 80

Rodenticides, 80

Fumigants, 80

Metals, 81

Solvents, 81

Animal and Insect Toxins, 81

Plant Toxins, 82

Reactive Hazards, 82

Unstable Substances, 82

Monomers, 83

Organic Peroxides , 83

Water Reactives, 84

Alkali Metals, 84

Alkaline Earth Metals, 84

Hydrides, 84

Carbides, 85

Nitrides, 85

Phosphides, 85

Inorganic Chlorides, 85

Peroxides, 85

Other Significant Water Reactives, 86

Air Reactives, 86

Biological Hazards, 86

Considerable Sources of Biohazards, 87

Blood and Body Fluids, 87

Animals, 88

Dusts and Spores, 88

Radiation Hazards, 88

Non-Ionizing Radiation Hazards, 89

Ionizing Radiation Hazards, 89

Exposure to Radiation, 89

Acute Exposures to Radiation, 90

Chronic Exposures to Radiation, 90

4   HAZARDS CHARACTERIZATION AND SITE EVALUATION   91

The Professional, 92

Site Categorization, 92

Someone Must Take Charge, 92

Routine Site Categorization Activities, 93

Off-Site Survey, Audit, and Evaluation Activities, 93

On-Site Survey, Audit, and Evaluation Activities, 94

Reconnaissance Personnel, 94

Second Entry Personnel, 95

Value of Documentation, 95

Acceptable Site Characterization Documents, 96

Site Sampling - Personal Sampling, 96

Sampling Techniques, 97

Air Sampling, 98

About Air Monitoring Equipment, 98

Oxygen and Combustible Gas Indicators (CGIs), 98

Toxic Atmosphere Indicat, 99

Photo-Ionization Detectors (PIDs), 99

Flame-Ionization Detectors (FIDs), 100

Aerosol Monitoring Devices, 100

Radiation Monitoring Devices, 101

Materials Sampling, 101

Sampling Plan, 102

Sample Collection Specifics, 103

Sample Records, Chain of Custody (Also Called Chain of Evidence)103

Risk Assessment, 104

Elements of Risk Assessment, 104

Environmental Health and Safety Site Hazards Evaluation Checklist, 106

Hazardous Materials Communication Site Evaluation Checklist, 111

Respiratory Protection Hazards Evaluation Checklist for APR Respirator, 115

Confined Space Entry Hazard Evaluation Checklist, 116

Ventilation Hazards Evaluation Checklist, 117

5 HAZARDOUS MATERIALS EMERGENCIES   **119**

The Best Time to Plan for an Emergency Is Before One Occurs, 119

About HAZWOPER, 121

Emergency Planning, 122

Contingency, 123

Elements of Incident Response, 123

Correlation Between Emergency Action and Emergency Response
Plans, 123

Emergency Response (ER) Plans, 124

The Emergency Response Triad of Importance, 124

Site-Specific Emergency Response Plans, 125

Emergency Personnel, 125

On-Site Personnel, 126

On-Site Emergency Response Team Leadership, 126

Incident Commander (IC) or Senior Response Official (SRO), 126

Incident Closure, 128

About Emergency Responders: On-Site Emergency Team Training, 129

Community emergency teams, 129

Classification and Duties of Emergency Response Personnel, 130

First Responder Awareness Level, 130

First Responder Operations Level, 130

First Responder Technician Level, 131

First Responder Specialist Level, 131

Incident Commander or Senior Response Official, 132

Incident Management, 133

Incident Commander, 135

Incident Response Procedures, 135

Guide 1.0 General HAZMAT Incident Response, 136

Guide 2.0 Response Guidelines, 137

   Personal Protection, 137

   Additional Considerations, 137

Guide 2.2 Sodium Hydroxide, Caustic Soda (50%), 138

   Personal Protection, 138

   Additional Considerations, 138

Guide 2.3 Diesel Fuel, Fuel Oil #2, 139

   Personal Protection, 139

   Additional Considerations, 139

Guide 2.4 Epichlorohydrin, 140

   Personal Protection, 140

   Additional Considerations, 140

Guide 2.5 Ferric Chloride Solutions, 141

   Personal Protection, 141

   Additional Considerations, 141

Guide 2.6 Gasolines, 142

   Personal Protection, 142

   Additional Considerations, 142

Guide 2.7 Hydrochloric Acid (33% solution), 143

   Personal Protection, 143

   Additional Considerations, 143

Guide 2.8 Hydrogen Peroxide (35% solution), 144

   Personal Protection, 144

   Additional Considerations, 144

Guide 2.9 Phosphorus oxychloride, 145

   Personal Protection, 145

   Additional Considerations, 145

Guide 2.10 Propylene Oxide, 146

Personal Protection, 146

Additional Considerations, 146

Guide 2.11 Sodium Hypochlorite, 147

Personal Protection, 147

Additional Considerations, 147

Guide 2.12 Sodium Bisulfite (40% Solution), 148

Personal Protection, 148

Additional Considerations, 148

Guide 2.13 Sulfuric Acid (93%), 149

Personal Protection, 149

Additional Considerations, 149

Appendix, 150

Sample Emergency Plans, 150

Definition of Emergency, 150

General Procedure, 150

Responsibilities Under This Plan, 150

Evacuation, 154

Supervisors, 154

Any Company Employees and Contractors, 154

1A. Emergency Procedures, 155

1A.2 Emergency Evacuation Drills, 156

In-Company Emergency Response Team, 156

1B. Emergency Procedures, 156

1B.2 General Chemical Spill Response Procedure, 157

1B.3 General Response to an Earthquake, 157

1B.4 General Response to Tornado, 158

1B.5 Emergency Notification Procedures, 158

Death or Multiple Injuries, 158

Hazardous Substance Spills or Releases, 159

Threats of Terrorism, 159

Training and Participation, 160

6   CORROSIVE SUBSTANCES   **161**

What Are Corrosive Hazards?, 161

  About Corrosive Liquids, 162

  About Corrosive Solids, 162

  About Corrosive Gases, 163

  Use of Corrosive Materials, 163

Protection from Corrosive Hazards, 164

  Ventilation, 164

  Personal Protection, 165

  Corrosive Hazard Protective Measure Checklist, 166

  Corrosive Air Contaminants, 168

  Health Effects from Corrosive Substances, 168

  Monitoring Potentials for Corrosive Exposure, 169

Safe Handling of Corrosives, 169

  Common Sense Procedures for Handling Corrosives, 170

  Corrosive Substance Storage, 171

  Bulk Storage of Corrosive Substances, 172

Emergency Planning and Spill Response for Corrosive Substances, 173

  Sample Guide for: Sodium Hydroxide, Caustic Soda (50%), 174

    Personal Protection, 174

    Additional Considerations, 174

  Sample Guide for: Sulfuric Acid (93%), 175

    Personal Protection, 175

    Additional Considerations, 175

  Acids, Anhydrides and Alkali, 176

    Safety Considerations, 176

    Health Considerations, 176

    First Aid for Corrosive Exposures, 177

      First Aid SOPs for Corrosive Exposure, 177

7   IGNITABLE SUBSTANCES   **183**

Characteristics of Ignitable Hazardous Substances, 183

Flammable and Combustible Liquids, 183

Other Considerations for Ignitable Liquids, 184

Ignitable Liquid Hazards Considerations, 185

Ignitable Liquid Specifics , 185

   Combustible Liquids, 185

   Flammable Liquids, 190

Ignitable Liquid Storage, 190

   Storage Specifics, 190

      Should plastic containers be used to store ignitable liquids?, 191

      Criteria for Ignitable Liquid Storage Cabinets, 191

      Ignitable Substance "Authorized" Storage Rooms, 191

Storage Room Rating and Capacity, 192

   Electrical Wiring, 193

   Ventilation and Air Quality, 193

   Solvent Room Storage, 193

   Damaged Containers or Leakage, 194

   General Purpose Public Warehouses, 194

Ignitable Substance Safety, 195

   Storage Reminder, 195

   Containers and Cabinets, 195

   Safe Handling Reminder, 196

Ignitable Solids , 196

   Catalytic Ignition, 197

   Pyrophoric Substances, 197

   Another Thought About Ignitable Solids, 197

About Dusts, 198

About Fumes, 198

About Vapors, 198

Solid, Liquid, or Gas, 199

About Mist, 199

About Ignitability and Explosives, 199

High Explosives, 200

Low Explosives, 200

Energies of Ignitability, 200

Decomposition, 200

Hazardous Polymerization, 201

8 HUMAN HEALTH HAZARDS   **203**

Paradigms of Human Health Hazards, 203

Characterization and Assignment of Health Hazard Status, 204

  Category 1—Urgent Public Health Hazards, 205

  Category 2—Public Health Hazard, 205

  Category 3—Indeterminate Public Health Hazard, 206

Safety and Health Considerations, 206

Lingering Community and Industrial Health Concerns, 210

  What About Mercury (Hg)?, 210

What about Chromium (Cr) Exposure?, 211

Environmental Health Concerns—Protect the Children, 212

  Asbestos-Induced Cancer, 212

  Asthma Concerns, 214

  Toxic Chemicals, Biocides, and Pesticides, 214

  Radon Gas Hazards, 215

  Tobacco Smoke, 215

  Protect Children from Lead, 215

  Hazards from Ultraviolet (UV) Radiation, 216

  Water Contamination, 216

9 BIOLOGICAL HAZARDS   **239**

Characterization of Biohazards, 239

  Further Characterization, 240

  Affected Populations, 240

Occupational Exposure to Biohazards, 240

  Virus Exposures, 243

  Bacterial Exposures, 243

Fungal Exposures, 244

Pathogenic Parasites, 244

Selected Biohazard Specifics, 245

Anthrax, 245

Influenza vs. Avian Flu, 246

Avian "Bird Virus" Genetic Variation, 247

Blood Borne Pathogens, 247

AIDS, 248

Botulism & Genus Clostridium, 249

Food Borne Diseases (Food Poisoning), 250

Hantavirus, 251

Legionnaires Disease, 251

Plague, 252

Smallpox, 252

Tularemia, 253

Viral Hemorrhagic Fevers (VHF), 253

10   DISASTER SITE WORK   **255**

Disaster, 256

Putting Disaster Sites Back to Normal, 256

Disaster Site Workers—Individual Preparation, 256

What Happens When Your Work Comes to an End?, 257

Disaster Site Cleanup - Abatement of Waterborne Biological Hazards, 258

Water Damage Classifications, 259

Type A.1 Clean Water, 259

Type A.2 Contaminated Water, 259

Type B Hazardous Water, 260

Microbial Contamination Characterization, 260

Initial Evaluations, 261

Follow Safe & Prudent Work Practices, 262

Site Inspection Checklists for Moisture Sources, Biohazards, and Other Hazards, 263

Personal Protective Equipment for Site Workers, 265

Classification, 265

OSHA Level D Standard Worker Protection Equipment, 265

OSHA Respiratory Protection Standard Guidelines, 265

OSHA Level C Total Coverage PPE, Air Purified Respirator, 266

OSHA Level B Total Coverage PPE, Supplied Air Respirator, 266

OSHA Level A Total Encapsulation Personal Protection, 266

Materials & Supplies, 267

Cleaners and Disinfectants, 267

Equipment and Tools, 268

Air Moving Equipment, 269

Dehumidifiers, 269

Other Drying Equipment, 269

Detection and Monitoring Equipment, 269

Microbiological Testing, 270

Mold Sampling, 271

Typical Sampling Methods Conducted for Direct Examination, 271

About Sampling, 271

Locations for Air Samples, 272

Lab Reports, 272

Interpretation of Microbial Testing, 273

Other Situations, 273

Take a Common Sense Approach, 274

Biohazard Abatement Plans Include, 274

Considerations for Water Damage Evaluators, Inspectors, and Project Managers, 275

Type A.1 Clean Water Intrusion, 276

Type A.2 or Type B Contaminated Water Intrusion, 276

Levels of Contamination, 277

Water Damage Mitigation Checklists, 277

Water Damage Clean up Examples, 279

Clean Up Type A.2 & Type B Water Damage, 280

Mold & Spore-Forming Biohazard Contamination Abatement Plans, 281

Containment Considerations, 282

Cleaning Considerations, 284

Level 1 Contamination Abatement Plans, 284

Level 2 Contamination Abatement Plans, 285

Level 3 Contamination Abatement Plans, 286

Level 4 Contamination Abatement Plans, 287

Level 5 & Level 6 HVAC Contamination Abatement Guidelines, 287

 Example Sequence for Cleaning and Decontaminating HVAC Systems, 288

References, 289

11 **CHARACTERIZATION OF CBRNE TERRORIST THREATS AND
    WEAPONS OF MASS DESTRUCTION (WMD)**   293

Disaster and Terrorism, 293

Considerations for Emergency Responders and Disaster Site Workers, 294

What are CBRNE Agents?, 294

 Chemical Characterization for CBRNE, 296

 Sources of Chemical CBRNE May Include, 296

TICS - Common Toxic Industrial Chemicals, 296

 The Release of Nerve Agents, 298

 Blister Agents, 298

 Blood Agents, 299

  Warning Signs for the Presence of Blood Agents, 299

 Choking - Pulmonary Agents, 299

  Warning Signs for the Presence of Choking Agents, 299

 Irritants—Substances Used in Riot Control, 300

  Warning Signs for the Presence of Irritating Agents, 300

Biological Characterization for CBRNE, 300

 Significant Biological Disaster—Is It Imminent?, 301

 How Are Biological Agents Detected, Confirmed, and Quantified?, 301

 U.S. Centers for Disease Control and Prevention - Characterize Potential
 Biological Hazard Agents, 302

 Category A Biohazards, 302

 Category B Biohazards, 302

 Category C Biohazards, 302

Characterization of Chemical & Biological (CBRNE) Agents, 303

Toxic Industrial Chemicals (TICS), 303

Biological Hazards, 308

Characterization of Radiological, Nuclear, and Explosive CBRNE Agents, 308

Radiological Dispersion Devices "Dirty Bombs", 315

What are Radiological Sources?, 315

How Can Dangerous Radiological Products be Obtained Illegally?, 316

What is the Value of Using Radiological Dispersal Devices for Terrorism?, 316

Considerations about Radiological Agents, 316

Nuclear Bombs and Explosives, 319

Common Nuclear Weapons That Have Been Developed, 320

Characteristics of Nuclear Bombs, 320

Dirty Bomb—Considered to be a Nuclear Weapon? No., 320

Modern Thermonuclear Weapons, 321

The Neutron Bomb, an Enhanced Radiation Weapon, 321

What Happens During a Nuclear Explosion?, 321

Four Characterizations of Energy Released During a Nuclear Event, 322

Conventional Explosives vs. Nuclear Explosives, 322

Blast Effects or Air Bursts, 322

Electromagnetic Thermal Radiation (ETR), 323

Electromagnetic Pulse (EMP), 323

Radiation, 323

Nuclear Fallout, 324

Conclusion, 324

Acknowledgement, 325

GLOSSARY 327

INDEX 345

Chapter 1

REGULATORY CONTROL OF DANGEROUS SUBSTANCES

An issue that should be driven by common sense involves regulatory definitions of hazardous materials, hazardous substances, and hazardous wastes. Let's deal with this issue now. In previous years, the regulatory agencies, special interest groups, and various environmental health and safety organizations have developed and used definitions for dangerous substances to meet specific "organizational" interests. For instance, the U.S. Department of Transportation (DOT) calls a dangerous substance "a hazardous substance" if the amount in a container requires regulatory reporting whenever it is released to the environment. The quantity of the hazardous substance is called "Reportable Quantity" or RQ. The same dangerous substance is also called "hazardous material" if the quantity in a container is less than the RQ. By the way, a listing of RQs is found in the DOT regulation 49CFR170-175, located at the end of a table found in section 172.101. Also the U.S. Environmental Protection Agency (EPA) maintains the RQ listing within their publication of the EPA List of Lists. The U.S. Occupational Safety and Health Administration (OSHA) may call a dangerous substance hazardous material or hazardous substance without many qualifiers, depending upon the publication. The International Civil Aviation Organization and associated organizations call hazardous materials "dangerous goods." For the purposes of this publication, let's call hazardous material any substance that can harm people or the environment. Let's also call this same stuff hazardous waste, whenever it loses value and must be disposed. Hazardous materials are dangerous substances that include chemicals and biohazards.

Because hazardous materials are of concern to the world community, it must be emphasized that they are closely monitored, more now than ever before. In the United

Figure 1.1 Hazmat regulations.

States, regulatory checks and balances have been established to help keep our workplaces safe, our environment clean, and our living conditions healthy. The food that we eat and the water that we drink are considered to be the best in the world. In this regard, regulatory controls have been valuable, even though they have increased costs of doing business.

The following agencies and organizations are involved in controlling the use of hazardous materials:

- INTERNATIONAL
  The Centers for Disease Control (CDC)
  International Civil Aviation Organization (ICAO)
  International Air Transport Agency (IATA)
  World Health Organization (WHO)
  Military Organizations of the United States and Allies
- U.S. FEDERAL AGENCIES
  Environmental Protection Agency (EPA)
  Occupational Safety & Health Administration (OSHA)
  Department of Transportation (DOT)

Federal Department of Health and Human Services
Federal Emergency Management Agency (FEMA)
Department of Energy (DOE)
National Institute of Occupational Safety (NIOSH)
National Fire Protection Agency (NFPA)
Federal Bureau of Investigation (FBI)
Central Intelligence Agency (CIA)
National Aerospace Administration (NASA)
- U.S. STATE AGENCIES
State Emergency Response Commission
State Environmental Agencies
State Health Agencies
State Emergency Services
State Police & Associations
State Fire Administration & Associates
- CITY & LOCAL AGENCIES
City & County Health Departments
Water Quality Department (Drinking Water & Water Treatment)
Department of Air Quality
Department of Waste Management
Emergency Services (LEPC, Fire, Emergency Medical, Law Enforcement ...)

REGULATORY CONTROL–THE CODE OF FEDERAL REGULATIONS

Regulatory control of hazardous materials that are produced and used in the United States begins with the Code of Federal Regulations. Hundreds of federal regulatory standards have been implemented over the past 30+ years. The most common hazardous materials regulations that impact industry and the community include:

- U.S. Environmental Protection Agency: Family of Standards = 40 CFR
- U.S. Occupational Safety and Health Administration: Family of Standards = 29 CFR
- U.S. Department of Transportation: Family of Standards = 49 CFR

NOTE: Each state and municipality also has developed written standards that deal with dangerous materials. Each has a unique coding system that stems back to the CFR's. States and municipalities are authorized to regulate dangerous substances from perspectives that can be more stringent than federal standards. Please remember "the most stringent standards apply."

ENVIRONMENTAL PROTECTION AGENCY (EPA)

The mission of the EPA focuses upon improving and preserving the quality of the environment. The EPA focus is both national and global, for protecting human health and the productivity of natural resources for which our lives depend. The span of

FOCUS AREAS OF WATER POLLUTION CONTROL

1. Industrial discharges: To establish permit programs in order to monitor sources of industrial pollution.
2. Sewage treatment methods: Standard methods and procedures acceptable for treating sewage. Publicly Owned Treatment Works (POTW) are closely monitored through EPA sponsored, state programs.
   - Acceptable only by using the best available treatment technologies.
   - States and local governments are responsible for ensuring proper sewer management.
3. Water run off: Reduction of "non-point sources".
   - Storm Water Permits are required for properties having potentials to cause pollution from storm water releases. Land owners are required to correct potential causes of pollution from their properties.
4. Spill Prevention Control and Countermeasure Plans (SPCC)
   Any operation that stores significant amounts of petroleum products and other substances are subject to:
   - Reporting of spills of oil and hazardous materials.
   - Possessing written and functional spill prevention plans.
   - Having a spill response and clean up program in place.
   - Are liable for clean-up costs.
5. Protection of U.S. Wetlands
   - Swamps and wetlands are to be maintained as valuable resources.
   - Have mechanisms in place for water purification, flooding, and erosion control
   - Preserve natural habitats for birds, fish, and wildlife
   - U.S. wetlands are permitted and regulated by Army Corps of Engineers.
   - EPA has the power to veto any development projects that may endanger national wetlands.

Table 1.1 Focus areas of water pollution control.

EPA coverage includes land, air, water and more recently, indoor air quality. The EPA's mission is:

- Enforcing U.S. environmental laws
- Integrating environmental protection with economic growth
- Evaluating potentials of environmental impact
- Using available technologies to remove or control environmental risks
- Stimulate public interests by showing benefits of environmental management
  1. Manage waste generation and reduce costs by reducing waste generation sources
  2. Encourage recycling of waste materials so resources can be reused and not disposed
  3. Re-use waste products and bi-products by providing them to others that have a use for them
  4. Reducing and eventually preventing pollution through hazard reduction, improved process efficiency, and by promoting energy conservation
- Promote national ethics for environmental stewardship

I. EPA: Clean Water Act

The objective is to restore the integrity of the waters of the United States by prohibiting the discharge of hazardous wastes and toxicants in the nation's waterways. Of prime focus is to govern pollution sources by standards of the National Pollution Discharge Elimination System (NPDES). This is done by seeking identifiable sources, called point sources and by identifying indirect pollution sources, called non-point sources.

II. Clean Air Act (CAA)

Stringent amendments were implemented during the early 1990's. However clean air regulations began as the Air Pollution Control Act of 1955. Objectives of the Clean Air Act include setting air quality standards for communities across the nation and to establish minimum acceptable emission tolerances from all identifiable sources. Major efforts have been made to list and regulate all serious industrial air pollutants.

National Ambient Air Quality Standards (NAAQS)
EPA has set standards for maximum allowable atmospheric concentrations of emissions that are determined hazardous to human health. These emissions include:

- Particulate matter
- Asbestos

- Carbon monoxide
- Nitrogen oxide
- Hydrocarbons
- Lead
- Ozone
- Photochemical oxidants

National Emission Standards For Hazardous Air Pollutants
Hazardous materials regulated under this section of the Clean Air Standards include:

- Asbestos
- Mercury
- Benzene
- Radio nuclides
- Beryllium
- Coke oven emissions
- Vinyl Chloride
- Others

Major Provisions of the Clean Air Act
Since 1990 several works have been completed to reduce airborne hazardous materials. They include regulating hundreds of sources of toxic industrial pollutants such as emission sources generating 10 tons/yr of any one hazardous air pollutant (HAP). Emission sources generating 25 tons/yr of combined volatile organic compounds are also known as (VOCs) and HAPS. EPA implemented applications for Maximum Achievable Control Technologies (MACTS) for new and existing hazardous air pollutant sources:

- Controls required in place before allowing new construction of operations having pollution potentials
- Controls required to be installed as existing sources are modified
- EPA completed a study that listed risks of air toxins after MACT rules are in place. The focus of the study was to check if controls actually have reduced aerosol-related illness (began in 1996, study not complete).
- Certain potential sources of air pollution were required to prepare risk assessments and management plans in the event of emergency releases of hazardous materials. Potentials for acid rain were reduced by limiting emissions of sulfur dioxide and nitrogen oxide, especially from fuel combustion sources.
- Allowances and incentives were given to companies that voluntarily reduced emissions, thus giving them the ability to sell their allowances on the open market. If a company was licensed to release 10 tons of emissions, and they maintained them for example to 5 tons, they were allowed to sell 5 tons of emission potentials to another company. Did that reduce pollutant emissions? Probably not.

- EPA implemented legislation to phase out chlorinated fluorocarbons (CFC's) and listed other substances that are believed to harm the ozone layer that surrounds the earth. Because of that action, aerosol sprays now contain flammable and toxic substances that may include propane, isobutene, alcohols, methylacrylate, propylene glycol, and similar chemistries. Remember Grandma warned us not to smoke in the outhouse? The same is true today about making sparks while applying deodorant. Issues relating to chlorofluorocarbons are still argued in a number of scientific arenas.
- Federal, state, and city air management agencies have increased civil and criminal penalties for violations of the Clean Air Act. Most violations may occur for less than one day and enforcement actions may begin.

III. Safe Drinking Water Act (SDWA)

For many years, the quality of drinking water varied significantly in cities across the United States. Major improvements have been made over the past 20 years. Not every city or town has perfect water, but epidemiology and toxicology studies have shown marked improvements. The prime objective of the SDWA is to provide quality drinking water for all human beings.

Air and water are the most abundant and essential resources needed for human survival. Fifty percent of drinking water comes from surface waters; 50% is extracted from wells and springs.

The SDWA establishes guidelines for Underground Injection Control for permitted "deep injection well" hazardous waste disposal. The Lead Contamination Control Act of 1988 prohibits using dissolvable lead in any pipe, solder, or flux used in public water systems. The Safe Drinking Water Act is primarily enforced at the state and local levels. EPA has power of enforcement through permitting and inspection.

Public Water Management

Local water districts are required to notify citizens if contaminant levels do not meet federal standards. Citizens may bring about civil lawsuits if water supplies become contaminated and are not treated.

IV. Hazardous Waste Management Resource Conservation And Recovery Act (RCRA)

(EPA Regulation: 40 CFR 260 -268)
The objective of the RCRA regulatory program is to protect natural and other resources of the United States by properly handling, storing, and disposing hazardous wastes.

A. Major Elements Of RCRA Regulations

The concept of "Cradle to Grave Responsibility" is that generators of hazardous wastes are liable for the wastes until the wastes or the generator no longer exists. In this regard, regulated waste generators should not use "shifty Sam and his brother-in-law" to illegally dispose any waste product. Hazardous wastes are closely tracked and monitored. Strict management is required for hazardous waste from generation thru process life to waste disposal by treatment or landfill.

Hazardous waste management control mechanisms focus upon specific directives to protect human health and the environment. They promote waste minimization standards and encourage principles for conserving natural resources.

B. Key RCRA Definitions

Hazardous Wastes: Hazardous materials without commercial value based upon physical properties, characteristics and the ability to pose an unreasonable risk to life, health, environment, and property.

Waste Generator: Any person that produces hazardous waste that is identified or listed by EPA, state, or local regulatory standards.

Treatment, Storage and Disposal Facility (TSDF): Licensed facilities that handle, treat, and dispose hazardous wastes.

Hazardous Waste Manifest: A standard document that is completed by the hazardous waste generator, the hazardous waste transporter and the treatment, storage and disposal facility. The document requires eventual routing to the state(s) hazardous waste regulatory agencies for tracking and monitoring. Other accompanying documents may include Land Ban Disposal Declarations, Waste Profiles, and other documents.

C. Hazardous Waste Identification

EPA has designated hundreds of wastes, as hazardous wastes, based upon known and tested characteristics. They are found in at least three lists shown in 40 CFR 261:

1. Non-specific Source Wastes (261.31)
2. Specific Source Wastes (261.32)
3. Commercial Products (261.33)

Also, documentation for record keeping, manifesting for shipment and tracking must include:

a. EPA and or State Hazardous Waste Numbers
b. EPA and or State listed hazard codes as they correspond to hazard characteristics
c. Department of Transportation (DOT) shipping criteria as listed in 49 CFR 170 – 175, which includes:
   - Proper waste shipping name, Hazard classification, United Nations or North American hazardous material number, Packaging group number
   - EPA Waste Codes (in some cases state codes also apply)
   - Reportable Quantity (RQ) listing as applicable
   - North American Emergency Response Guidebook Number

D. Reportable Quantities (RQ)

Directed by the Superfund Acts (CERCLA & SARA) the Environmental Protection Agency designates specific amounts of hazardous substances as being reportable when spilled, released, or involved in emergency situations. These substances are listed in the Department of Transportation tables found in 49 CFR 172.101. The amounts listed are based upon the total weight of a reportable hazardous material in an individual container.

Spills and releases of R.Q. hazardous materials (RQ amounts are called hazardous substances) to the environment, must be reported to designated emergency response and regulatory agencies. Agencies that require reports may include: The Local Emergency Planning Committee, Fire Departments, Law Enforcement Agencies, Water Treatment Facilities (POTW), City Air Department, Water and Health Departments, State Environmental and Health Departments, the National Response Center, and the Federal Environmental Protection Agency.

E. Characterization of Hazardous Wastes

Hazardous wastes must be properly managed and properly disposed. How do we know if a substance is a hazardous waste? Look in 40 CFR, Section 261 and determine if it meets listed characteristics such as:

- Ignitable: Solids and gaseous materials that readily ignite and burn. Also, liquids that become air at 140 degrees Fahrenheit or less, and will burn if an ignition source is present (flashpoint).
- Reactive: Substances that react with other materials and produce hazardous conditions, especially if disposed in a city land fill.
- Corrosive: Substances that will destroy human and animal flesh, plants and will dissolve metals. EPA lists corrosive hazardous materials as products demonstrating physical corrosive properties and pH criteria. Such as, acids having pH measurements of 2 or less and alkalis having pH measurements of 12.5 or higher.
- Toxic: The largest listing of hazardous wastes in Section 261. Basically anything that can cause a health concern may meet toxic criteria. Caution, keep an eye open for materials that meet toxic characteristic leaching procedure (TCLP) criteria, as described in Section 261. If present, several listed substances in small quantities, may qualify as TCLP hazardous wastes. The TCLP tests to determine if regulated materials, when exposed to water, will allow regulated quantities of the hazardous material to leach into water. This simulates wastes that may be placed in a public landfill and determines what will happen when rain waters soak the wastes.

Note: Certain substances may be hazardous and are not regulated by RCRA because they are regulated by another agency. These could include household wastes, municipal recovery wastes, agricultural wastes, mining overburden, radioactive wastes, and others.

F. Hazardous Waste Management Requirement

Regulatory Agency Notification

When an organization generates hazardous waste, it is responsible for that waste until it is completely treated, "cradle to grave." The waste must be properly treated, stored, and disposed of on site, or it must be properly transported to another location for the same purpose. EPA and most states require notification by submitting required documentation to them before hazardous waste management activities begin.

Hazardous Waste Permits (40 CFR 270)

If hazardous waste treatment and disposal is performed on site, generators must apply for and be granted permits from EPA and from applicable state regulatory agencies. Short term, 90 day permits may be granted to allow generators to accumulate minimum quantities for transportation.

Hazardous Waste Tracking

If hazardous wastes are to be transported off site, the transporter must be licensed and registered with EPA. An EPA Transporter Number must be issued to the transporter. Shipment of hazardous wastes must be documented on a uniform hazardous waste manifest and generator identification numbers and transporter identification numbers must be included on the manifest for regulatory tracking. Generators are required to prepare Land Ban Statements and include them with shipping documents.

Hazardous Waste Reporting (40 CFR 262-265)

Generators, transporters, and TSD Facilities are covered by hazardous waste reporting requirements. These reports may include quarterly waste activity reports, biannual reports, accidental release reports, and community right to know reports.

Record Keeping (262-265)

Hazardous waste shipping and tracking records are required to be maintained for at least three years. Copies of Land Ban Statements are required to be maintained for longer periods of time (seven years in some cases).

   Note: It is recommended that anyone involved in hazardous waste management keep up-to-date on regulatory compliance requirements. There may be various differences in management requirements from state to state. In all cases, the most stringent laws apply. Please review the appendices of this publication and note examples of hazardous waste compliance inspection checklists from various states.

V. Superfund Acts

Prior to 1980, there were many communities across the nation that experienced wide varieties of environmental disasters related to irresponsible methods of disposing waste hazardous substances. Many hazardous sites existed and still do exist that were uncontrolled and funds were not available to clean up the pollution. Congress acted to handle this problem. Waste sites were prioritized into the National Priorities List (NPL), as money becomes available they are decontaminated.

CERCLA

The Comprehensive Environmental Response Compensation and Liability Acts were established after decades of environmental disasters occurred. CERCLA was provided to generate funds and action for hazardous waste site clean up. It became known as Superfund. The objective of CERCLA was to empower the federal government to comprehensively respond to environmental disasters and to pursue potentially responsible parties.

Congress established two trust funds under CERCLA. One is the Hazardous Substance Response Fund designed to cover response costs of emergency "accidental" spills. The Post Closure Liability Fund is designed to pay clean up costs for abandoned and inactive waste sites that are listed on the National Priorities List (NPL).

Under CERCLA, four kinds of costs are documented and can be recovered:

1. State or federal costs resulting from site clean up
2. Costs incurred by private contractors
3. Damaged resources owned or controlled by public entities
4. Costs related to health assessments and remedial care

Significant CERCLA considerations include:

1. Responsible party notification of reportable quantity releases to National Response Centers
2. Record keeping of all reportable quantity (RQ) releases
3. Creation of the National Response Team to respond to large scale chemical releases
   - The NRT can involve as many as 15 federal agencies.
   - The NRT developed a National Contingency Plan.

Superfund Amendment And Reauthorization Act (SARA)

When legal actions were taken to clean up contaminated waste sites across the United States, Congress acquired money to pay for cleanup operations. In a short amount of time, it was found that more money was needed to accomplish this goal. The money was not readily available. In 1986 actions occurred to amend and reauthorize the CERCLA program. As a result the following program titles emerged:

Title I

- To get more money for site clean up

- To establish the National Priorities List criteria which includes:
  a. Proximity of contamination to schools and communities
  b. Locations of near by natural resource.
  c. Severity and hazardous properties materials contamination
  d. Immediate consequences if site and surrounding areas are not decontaminated

Title II
This portion of SARA focused upon the workers right to know about potential exposures to harmful substances while performing duties at contaminated sites. In 1987 the OSHA Hazard Communication Standard was implemented to protect workers in all areas, (please see the OSHA Hazard Communication Standard 29 CFR 1910.1200).

VI. Emergency Planning and Community Right to Know (EPCRA)

The objective this regulation is to provide information to public emergency response agencies and to local governments regarding chemical hazards in the community. A key focus of the standard is to encourage emergency planning at all levels of the community and government. Prime incentives for EPCRA are to obtain information that can be used by emergency responders prior to responding to hazardous materials spills, releases, or other incidents. Also included is the requirement to reduce potentials for catastrophic results of incidents such as that seen in Bhopal India, as well as incidents that have occurred in the United States.

Emergency Planning and Notification
Governors of each state were required to establish a State Emergency Response Commission (SERC). Local Emergency Planning Committees (LEPC) were established to prepare community emergency response plans.
   Members of the LEPC include:

- Fire Department Officials
- Law Enforcement Agencies
- Health Department Officials
- Hospitals
- Transportation Organizations
- Public Works Officials
- Heads of Government
- News Media
- Industry Representatives

Reporting and Notification
Any site having an inventory of hazardous substances that are listed as "reportable," in quantities that are reportable in the EPA List of Lists, must submit applicable reports. All organizations having a significant inventory of hazardous substances are required to provide an inventory listing, or a full set of Material Safety Data Sheets (MSDS) to the LEPC. Refer to EPCRA Tier I and Tier II reporting and other related requirements for federal and state reporting of hazardous materials inventories and usage. Also note Section 313, Toxic Release Inventory reporting requirements shown on the EPA list of lists. The Community Right to Know Hotline, 1-800-535-0202, is a resource for additional information.

Comprehensive Emergency Response Plans
The Federal Emergency Management Administration (FEMA) is instrumental in assisting state and local governments in preparing emergency contingency plans. Chemical emergency contingency plans should provide the following information:
    Identify locations of hazardous substances:

- Designated transportation routes
- Establish procedures for notifying responders
- Establish standard operating procedures for responders
- Develop evacuation plans and procedures
- Describe how the severity of each incident is determined
- Establish methods of notifying the public
- Provide for the use of emergency equipment
- Describe how training will occur for response personnel
- Establish a frequency for conducting community emergency response drills and scenarios
- Designate community and facility coordinators

VII. Pollution Prevention Act

In 1990 the Pollution Prevention Act was passed by Congress. The highest priority of this act is to reduce pollution at each source. Key elements of the Act include prevention by replacing hazardous materials with suitable non-hazardous substitutes and maximizing production output using minimal amounts of raw ingredients and to promote efficient use of energy. Recycling and reusing materials is done rather than increasing pollution potentials by disposal. When materials must be disposed, using non-polluting treatment methods. Minimize wastes by every means possible. Dispose of hazardous wastes only as a last resort.

Pollution Prevention Ethic
The following are elements of the Pollution Prevention Ethic:

- When possible, don't create hazardous wastes
- Reduce amounts of hazardous materials used in processing
- Find ways to reuse hazardous materials in processes
- Recycle hazardous materials rather than having them disposed
- Return usable hazardous materials to the manufacturers
- Conserve energy resources whenever possible
- Reduce amounts of packaging where practical
- Foster a philosophy that having the most is not always the best policy

VIII. Occupational Safety and Health Act (OSH Act)

Key objectives of the Occupational Safety and Health Administration (OSHA) are to develop safety and health standards in the workplace, support education and consultation services and to enforce occupational safety and health standards.

Safety and Health Standards
OSHA standards are promulgated using histories of known tragedies, court actions, advents of new technologies, initiatives from within the agency, industry, and public petitions.

Who Is NIOSH?
National Institute for Occupational Safety and Health (NIOSH) is one of the research arms used by the U.S. Department of Health and Human Services. NIOSH performs tests and provides recommendations to OSHA that are used in developing safety and health standards.

OSHA Requirements for Employers
OSHA requires that employers provide safe workplaces for all employees, as directed by the General Duty Clause. OSHA has developed a hierarchy of controls that employers must comply with, which includes:

1. Identifying all workplace hazards
2. Reducing and eliminating workplace hazards by applying engineering controls
3. Communicating hazards information to all applicable employees
4. Provide personal protective equipment to employees

1. Examples of Considerations for Evaluating Safe Work Places
   a. Determining applications for using engineering controls:
   - Using the best available technologies
   - Proper installation of heating and ventilation systems
   - Proper lighting of the workplace
   - Building ergonomics into process methods and procedures

- Applying mechanical assistance where possible.

b. Considerations for assessing hazardous environments:
  - Site security management
  - Work-zone designations and site layout
  - Preventing unauthorized access to hazardous activities
  - Preventing exposures to hazards and hazardous conditions

c. Information and training requirements:
  - Conducting on the job training (onsite and offsite)
  - Establishing and enforcing safe standard operating procedures
  - Distributing information at safety meetings
  - Coordinating routine safety and health reminders at production meetings
  - Providing safety and health information during new employee orientations

d. Limiting employee exposures to hazards and hazardous conditions:
  - By conducting routine work place environmental monitoring
  - By building principles of industrial hygiene into all work place activities

e. Ensure that a mechanism is in place that will help employees be aware of work-place hazards:
  - By complying with the OSHA Hazard Communication Standard
  - Placards signs and labels: Danger, caution, instructional signs, hazardous substance signs, limits, warnings, product labels and using OSHA Hazard Communication Labels

f. Maintain safety and health records for:
  - Accidental occurrences
  - Near misses
  - Employee injuries
  - Lost time
  - Specific OSHA standard recordkeeping for applicable safety and health programs

g. Maintain employee medical programs such as:
  - Performing medical surveillance
  - Recording and following up on employee exposures, injuries, and illness

h. Having a program to ensure the proper use of personal protective equipment:
  - Requires official initial and update site evaluations
  - Proper selection and fit of PPE to meet employee needs
  - Employee training to properly use, care and maintain personal protective equipment

i. Develop and implement site emergency action and response programs:
  - Emergency action plans must be written and functional

- Emergency response plans must be written and functional
- Emergency action and response plans must interface with one another. All expected actions that employees are expected to perform must be precisely listed in the site emergency programs.

2. OSHA Enforcement
The objectives of OSHA enforcement are to check how worker safety and health standards are implemented into the work force and to determine if they are being properly followed. How is this done?

- By conducting inspections performed by federal and state inspectors
- By performing third party environmental, health and safety audits and evaluations of the work place
- By performing self audits using designated employees to routinely perform inspections
  a. OSHA will follow up on employee complaints
     Inspections conducted after employee requests
     Employee names are held confidential, when requested
  b. OSHA will follow up on state insurance records to evaluate routine occurrences of employee injuries

3. OSHA Hazard Communication Program
The OSHA Hazard Communications Standard, also called "The Workers' Right to Know," was established for protecting workers that work with and around hazardous substances. The OSHA HAZCOMM Standard is 29 CFR 1910.1200.

KEY ELEMENTS OF HAZCOMM

a. Material Data Sheets (MSDS)
- MSDS are provided by manufacturers of hazardous substances, at minimum on the first time a product containing hazardous materials is purchased and whenever a change is made to the product.
- Employees are required to read and understand MSDS information before working with any hazardous material.
- MSDS must be kept "readily available" in the workplace.
- Chemical listings and inventories must correlate with copies of MSDS.
- MSDS and chemical inventories must be kept up-to-date.
b. Hazardous Materials Labeling
- All containers must be marked after original containers are opened and chemicals are transferred to new containers.
- Show names of hazardous material

- List specific hazard(s) of the product
- List employee protection requirements while using the product
- List target organs that can be affected
- Show manufacturers name and other data

c. Hazard Communication Training
- Training is to be provided before employees begin working with hazardous products.
- Employees must understand risks and hazard potentials.
- Employees shall be trained in methods for safe handling and use of hazardous products.
- Workers shall be knowledgeable about protection requirements and use of PPE.
- Training shall inform employees about MSDS information, including where MSDS copies are located at the work place and how to find information within them.
- Hazard Communication training is an on-going program. Not to be given only annually. In fact, training is required each time a change in processing occurs, or whenever hazardous products are changed.

4. Informing Workers About Workplace Hazards
Employers are responsible to provide workers with information about workplace hazards. Employees are responsible to follow safe work practices and also to communicate hazard information to all applicable personnel. Workplace hazards are determined by performing site assessments and evaluations. Site assessments and evaluations should include the following hazard considerations. Please note that many of the following subjects will be covered in following sections of this book:

- The properties of hazardous materials used and stored on site
- Locations of hazardous operations and conditions
- Specific hazardous materials exposure potentials
- Specific fire and explosion potentials
- The existence of noise hazards
- Environmental temperature extremes and stress factors for each
- Electronic and magnetic hazards
- Microwaves and radio frequency hazards
- General and physical safety hazards that may exist throughout the site
- Equipment traffic hazards
- Specific equipment hazards
- Oxygen deficiency potentials
- Site contamination potentials

- Ionizing radiation hazards
- Non-ionizing radiation hazards
- Biological hazards
- Confined space work hazards
- Methods of controlling hazardous energy
- General industry safety and health considerations
- Contractor and construction safety and health considerations

Chapter 2

PRINCIPLES OF SAFETY

SAFE WORK PRACTICES

Anyone that deals with dangerous substances must be aware that safe work practices are required for survival. Someone once said, "There are not many old bold pilots," and it must also be said that, "There are not many old bold hazardous materials professionals." How do we determine if our modes of operation are safe? Well, common sense must first prevail. Unfortunately, courses in common sense are hard to find.

OSHA Self Evaluation

A great source that helps in this endeavor is the Federal Occupational Safety and Health Administration. Search the OSHA web site at www.OSHA.gov and note all the programs and guidance documents that are free for downloading. Please don't try to memorize all the OSHA and other regulatory references; save that for people who don't have enough work to keep themselves occupied. What we all need to do is learn where to look in the regulatory Internet sites. Most of us involved in safety and environmental consulting swim daily in the deep pools of regulatory information. Some people spend hundreds of dollars to buy canned programs that were developed from information that is found in regulatory web sites. Why do that? Get the information directly from its source; don't pay the middleman for easy-to-retrieve data. Let's take a look at some of the regulatory guidance materials and supplemental information that can help us to be better at conducting hazardous materials evaluations.

Regulatory Programs Pertaining to Hazardous Materials Work Sites

There are some basic HAZMAT regulatory areas that most safety and environmental professionals should know about. They include:

- Emergency action and evacuation, plans and programs (OSHA)
- Emergency Planning and Response (OSHA)
- Hazardous Materials and Waste Management (EPA/RCRA)
- Hazardous Waste Operations and Emergency Response (OSHA)
- Hazards Communication (OSHA)
- Blood Borne Pathogens (OSHA)
- First Aid (OSHA)
- Process Safety Management (OSHA)
- Fire Safety, Fire Suppression and Response (OSHA & EPCRA)
- Control of Hazardous Energy (lock out / tag out) (OSHA)
- Confined Space Entry, Permit Spaces and Confined Space Rescue (OSHA)
- Personal Protective Equipment Assessment and Certification (OSHA)
- OSHA Record Keeping (OSHA)
- Materials Handling and Storage (OSHA, DOT)
- Powered Industrial Trucks, Forklifts and other... (OSHA)
- Substance Specific Programs (lead, mercury and many more) (OSHA)
- General Duty Clause of the OSH Act (OSHA)

Other associated programs include:

- Department of Transportation Shipping and Handling of HAZMAT (DOT)
- International Civil Aviation Organization, Air Shipment of HAZMAT (DOT)
- Spill Contingency and Countermeasures for Regulated Materials Releases (EPA)
- Clean Water and Water Pollution Control Programs (EPA)
- Clean Air Act and Air Pollution Control Programs (EPA)
- HAZMAT Security and Anti-Terrorism Awareness (DOT/Homeland Security)
- RCRA Waste Management and Contingency (EPA)
- Otherwise Regulated Wastes, Special Wastes and Universal Wastes (EPA)

Don't forget that every state also has specific safety, health, and environmental regulations and most cities have ordinances that may be more stringent than the Federal programs listed above. The most stringent requirements apply. If you need regulation references, search the Internet for each regulatory program and respective agency listed above. We should now discuss some specifics that may impact safety

Figure 2.1   OSHA & CDC/NIOSH hazardous materials guidance
resources are readily available for all HAZMAT workers.

evaluations at some work sites, as well as assessments that are made to establish safe
work practices.

Considerations for Safe Work Practices

In every situation that involves hazardous conditions, someone must take charge.
People that coordinate work activities are often responsible to ensure that prescribed
work practices are safe. Site safety officers, safety, health, and environmental managers or whatever designated "safety and environmental professionals" may be
called are usually also responsible to ensure that safe practices are followed.
Regulatory programs should be properly spliced into Standard Operating Procedures
(SOP's). Hazards recognition and site assessments for potentially hazardous conditions are precursor activities for implementing proper standard operating procedures
and for ensuring that safe work practices can be followed.

What kinds of work practices require evaluation? They all do. Let's discuss some specific activities that some folks have overlooked before finalizing SOP's. Please note that each of the following considerations is driven by regulatory standards, and also note that these issues are laced into samples of site safety checklists shown later in this chapter.

COMMUNICATING HAZARDS INFORMATION

If dangerous materials are involved in work activities, naturally the OSHA Hazard Communication Program (HAZCOM) must prevail. What's involved in complete hazardous materials communication? In this regard, two perspectives should be addressed:

- Employee Right to Know (OSHA HAZCOM)
- Community Right to Know (EPA EPCRA)

Employee Right to Know

As mentioned in other portions of this publication, employees must be aware of what they are working with, and what can happen to them if they don't properly handle materials that can hurt them. Employees must be aware of dangerous properties related to hazardous materials before working with them. How many times do we hear news about workers that are seriously injured because they were not properly trained, or they overlooked hazardous conditions that arose from improperly handling dangerous materials? Don't forget about the "good ole boy" that was puffing on a cigarette while pumping gas into his pickup truck. This guy lucked out with this practice several times until the time that gasoline vapors exploded in the fuel tank he was filling, and he became a "crispy critter." Later in this chapter, Hazards Recognition, we will review key requirements of OSHA HAZCOM.

Community Right to Know

Several years ago in Kansas City, Missouri, six firemen responded to a trailer fire that was set as a prank by a group of high school kids. The trailer was parked on property owned by a construction company that performed work to widen Missouri highways. The contents of the burning trailer included ammonium nitrate, a fuel oil mixture that is used as an explosive. The firemen had an idea that the construction company worked on highways, but they did not know what was in the burning trailer. Explosives were not properly reported. The burning trailer and one parked nearby were not properly placarded. Both trailers exploded and all the firefighters perished. Today in Kansas City, as well as in many other American cities, all significant inventories of hazardous materials must be reported and locations of hazardous materials must be marked by posting NFPA 704 placards. Annual Tier II reporting is required by anyone having significant inventories of dangerous materials, irregardless of the storage location. Annually on March 1, reports must be sent to State Emergency Response Commissions (SERC), Local Emergency Planning Committees (LEPC) and local fire departments. We will discuss specifics of NFPA 704 placards later in this chapter.

WALKING AND WORKING SURFACES

When people trip, slip, or fall, they usually become injured and in far too many cases they are killed. Issues that must be considered to promote safe work practices involving walking and working surfaces include:

- Openings: Locations of openings and potential unsafe conditions that may result from openings within work areas.
- Stairs: Unsafe conditions should be considered that may relate to damaged or weak stairways, as well as weak or missing handrails.
- Ladders: Ladder safety is often overlooked in most industries. Ladders must be authorized, of proper strength, and routinely inspected.
- Scaffolds: Falls from scaffolds occur far too often. Proper assembly, design, and authorized use are issues that must included in evaluating safe work practices.

OVERHEAD AND UNDERGROUND UTILITIES

Drilling and digging operations must consider locations of hidden or overlooked utility installations. Locations of electrical wiring, natural gas plumbing, fresh water, sewer wastewater, and emergency fuel plumbing are serious considerations that must be included in evaluating safe work practices.

- Underground Considerations: Most states require that utility companies be consulted before digging operations are authorized.
- Overhead Considerations: Workers must always be aware of electrical power lines, locations of pipes as well as other utilities before starting work and must remain aware while activities occur.

ELECTRICAL SAFETY

Should electrical fans be located near operations that use volatile, flammable liquids or gases? The answer is no, if the fan or electrical device is not explosion-proof. There are other issues to consider for people that perform work on electrical systems. They include:

- Authorized, proper protective equipment
- Proper clearance when working on electrical equipment
- Are electrical workers properly trained and qualified?

TOOLS AND HEAVY EQUIPMENT

When we routinely operate equipment and become comfortable with our selected modes of operation, that's when we often have an accident. When we operate an automobile in stormy weather, we usually slow down and consider safe contingencies. What about that nice drive home when the sun is so perfect and the last thing on our mind is an accident? Unfortunately, some parents never return home safely to their children because the last thing on their mind was the first thing that happened. When we operate equipment we should always:

- Check that the equipment has been properly maintained.
- Inspect equipment before operation.
- Operate the equipment as was intended to be operated. Follow prescribed safe operating procedures.

PROTECTION FROM HAZARDOUS ENERGY

Equipment operators sometimes need to temporarily shut down normal operations in order to make adjustments and other reasons. In situations when multiple operations are occurring within a process, or maintenance and repairs are being made, hazardous energies must be controlled to prevent injury or death to people. What kinds of energies need evaluation and isolation?

- Electrical
- Pneumatic
- Explosive
- Hydraulic
- Steam
- Mechanical

Hazardous energy isolation methods are regulated by OSHA in the Lock Out and Tag Out Standard. Equipment and energy sources must be locked to prevent activation while people within danger zones. Sources of equipment activating energies shall be tagged in order communicate that the source is locked and isolated, until released by applicable parties.

LIFTING AND REPETITIVE MOTIONS

Improper lifting is the cause of most back injuries. It should be simple to prevent most back injuries by using our legs more efficiently. However, many people don't take that moment to think of using proper lifting methods before they begin moving a heavy object. Musculoskeletal disorders are not new, even though the term "ergonomics" is now a household word. Repetitive motion injuries and job-related aches and pains have been around a long time. However, regulatory and legal issues have raised a red flag for employers to notice. Emphasis must be made to engineer the work place to fit the employees that work in it. Employers must provide proper equipment and training that will enable workers to complete their assigned tasks without pain and injury. Employers should take advantage of ergonomics training information that is found in the OSHA web site; it's free.

The Federal Occupational Safety and Health Administration attempted to implement a mandatory standard that would help reduce ergonomic disorders in the work place. However, it became a voluntary standard because most industries in the United States would have suffered major costs to implement such as standard. It must be noted that all employees are covered by the OSH Act and the General Duty Clause. In this respect, all workers are guaranteed a safe work place, and it's the law.

Figure 2.2 HAZMAT employees that wear personal protective equipment must use proper procedures to prevent injury (OSHA PPE and Ergonomics Standards).

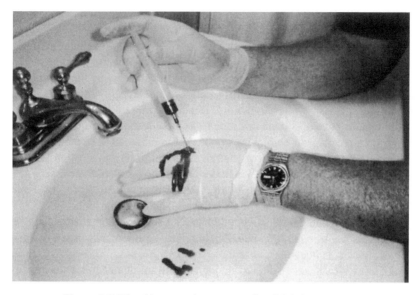

Figure 2.3 Blood borne pathogens, needle stick injury (DAS).

All employers are encouraged to develop applicable ergonomics programs to prevent ergonomics injuries. What's involved in ergonomics compliance?

Design work and the work place to meet the physical needs of the employee. The human body does not adapt to all working conditions. Provide training and obtain proper tools that will eliminate millions of dollars in injuries and lost time.

HUMAN BLOOD AND BODY FLUIDS

Less than thirty years ago, the human immunodeficiency-virus (HIV) was not prevalent. Most of us who earned degrees in microbiology and virology knew about weird diseases that caused the immune system to attack a person's own body (auto-immune syndrome). However we did not know of a micro-organism that stimulated etiology so perfectly that it caused the body's immune system to become retarded. In some circles HIV is thought to be man-made. Regardless of where it came from, AIDS is a prevalent fatal disease that is spreading across the globe today. Micro-organisms such as HIV and Hepatitis viruses are obligate parasites that live within the living cells of human body fluids. As long as the cells stay alive, so do the viruses. OSHA promulgated the Blood Borne Pathogens Standard to protect workers that may become exposed to pathogenic organisms found in human body fluids.

Safety considerations for all people that may contact human body fluids must include the following:

- Take universal precautions when dealing with human body fluids. In all cases, human body fluids should be considered contaminated with pathogens.
- Prevent exposure when providing first aid, rendering medical treatment, or decontaminating spills of body fluids by wearing proper personal protective equipment (refer to www.osha.gov and review all requirements relating to handling blood borne pathogens).

BIOHAZARDS

Blood borne pathogens are only a few microbiological health hazards that workers should consider in evaluating site safety. Anyone that shines a light into a pit or crawlspace and sees eyes looking back at them should consider that biohazards are awaiting them. As mentioned in Chapter 3, biological health hazards may be in the form of bacteria, viruses, fungi, and other forms of micro-organisms. Biohazards may exist within host cells of animals and plants, including blood borne pathogens as mentioned above. In October 2003, the Occupational Safety and Health Administration issued a guidance document for workers that perform mold abatement work. It is a standard that reflects training, standard operating procedures, cautions, and protection as directed by the OSHA HAZWOPER Standard for site workers. Thousands of workers have performed jobs that have exposed them to microbial hazards without knowing it. Whenever flood, fire, and water damage operations are performed, a Pandora's Box filled with chemical and microbial agents are awaiting unprotected workers.

HAZARDOUS MATERIALS SPILLS AND RELEASES

In the past anytime a substance was spilled or released to the environment, irregardless if it were hazardous or regulated, the person(s) responsible for the spillage were responsible to clean it up. In other situations, whenever there was a fire, flood, or another catastrophe, anyone nearby was enlisted to clean up and remediate the disaster area. In some cases, owners and operators would hire temporary "help," often vagrants or illegal immigrants, to clean up a hazardous incident area. At the end of the shift, the temporary workers (now exposure victims) would be paid cash and dropped off on the outskirts of town. What's wrong with that? Today acts such as these are considered criminal.

In 1990, Congress blessed a new law called HAZWOPER. This law or OSHA Standard means Hazardous Waste Operations and Emergency Response. This standard explicitly covers workers that may perform clean up operations such as:

- Controlled waste sites at waste treatment, storage and disposal sites
- Uncontrolled waste sites, such as areas of widespread contamination
- Industrial spill teams that deal with chemical and biological spills and releases
- Public and private emergency response teams that respond in a wide variety of ways
- Fire department personnel that stabilize incident areas
- Law enforcement personnel that respond to accidents, acts of violence, acts of crime, terrorist actions, forensics, clandestine drug scenes, and other sites
- Emergency medical personnel that support fire and law enforcement personnel operations
- Abatement and remediation contractors that abate chemical, biological, and electromagnetic contaminations in wide varieties of areas

Originally, the U.S. Environmental Protection Agency was tasked to regulate what is now called OSHA HAZWOPER, due to releases and potential releases of regulated materials. However, the focus of the HAZWOPER Standard covers worker safety in the above- mentioned areas. It must also be mentioned that the Hazardous Waste Operations and Emergency Response Standard also deals with interface activities that prevent and control environmental emergencies and ship-

Figure 2.4 Emergency response team decontamination of personnel wearing Level B personal protective equipment by personnel wearing Level C equipment. (HMTRI/CCCHST)

ping, handling, and storage of dangerous goods. The HAZWOPER Standard has such broad coverage in this regard that major areas of OSHA regulation are covered, as are areas of EPA regulation of environmental contamination and DOT regulation of all modes of hazardous materials emergencies during shipment. HAZWOPER activities will be presented in other chapters of this text.

RECOGNIZING HAZARDS

What is a hazard? It could be a substance, a situation, or condition that is capable of causing harm to people, property, and/or the environment. In regulatory, legal, and insurance perspectives, the scope of this definition could depend upon special interests. Again, common sense must prevail.

What is risk? A quantitative measure of risk could be based upon probability and potential severity of a hazard and its potential effect upon life, property, and/or the environment. Based upon the previous definitions, safety could be considered a judgment of a given risk as to whether it is acceptable as compared to perceived outcome(s) that may result from taking a given risk.

SITE SPECIFIC HEALTH AND SAFETY PLANS (HASP)

A measure of a safe work site should be more than quantitative data that indicates accidents and illnesses are less than the national average. Site conditions, standard operating procedures, identification, and control of hazard potentials are key elements to be included in recognizing job site hazards. Job safety analysis (JSA) activities should focus upon tasks that are performed at any given work site. JSA activities should include:

- Having the ability to recognize hazardous conditions
- Identification of tasks that demonstrate a potential for incident, injury, or illness for work site personnel
- Each step of every specific task should be listed and evaluated for hazard, risk, and safety.
- Standard operating procedures used to perform each task should include an evaluation for proper procedure, safe work practices, and overall protection of each employee.

Every work site contains potential for multiple hazards, especially when dangerous substances are involved. In most cases, 100% work place safety is very difficult to guarantee unless strict measures are taken to manage, control, monitor, and routinely evaluate mechanisms for controlling all unsafe conditions.

Process: Alkaline Cleaner
HAZARDOUS PRODUCT: Sodium Hydroxide

By Schmaltz Chemical

HAZARDS: Corrosive &Toxic Health Hazard

Destroys Human Skin, Tissues, and Metals

PR0TECTION: Ventilate, Avoid Contact, and Prevent Skin and Eye Exposure....

Wear: Rubber Gloves; Eye Protection; Splash Protection

FIRST AID: Wash affected area with cool clean water for 20 minutes; report exposure and injury to supervisor; seek immediate medical attention for injury.

TANK NUMBER: #42

Figure 2.5 OSHA chemical hazard communication.

RECOGNIZING HAZARDS RELATED TO DANGEROUS MATERIALS

Earlier in this chapter, we discussed communicating hazards presented by dangerous materials. We discussed the OSHA Hazard Communication Standard, the federal regulation that requires all employees to be able to recognize chemical hazards, know how to work safely with dangerous substances, and know how to recognize symptoms of exposure. At minimum, the OSHA Hazard Communication Standard, or "Employees' Right to Know Regulation" requires employers to have an active, functional program in place before any employee begins work with any dangerous material. The required elements of an active OSHA Hazard Communication Program are listed in the following section.

Chemical Hazard Communication Program

The following is required for an active and functional HAZCOMM program:

1. Responsibility: Someone must oversee and manage the site-specific program.

2. Written plan and program: If a safety program is not written as site-specific, it does not exist. In this regard, generic written plans and procedures are not acceptable unless they cover site-specific criteria that will guarantee communication of hazards information to all employees. What if all the employees on a given work site do not speak or understand the same language? The employer is responsible to ensure that all employees have all the proper information communicated to them so they will recognize work place hazards, understand how to work safely with dangerous or otherwise hazardous materials, and are knowledgeable about every element covered by the HAZCOMM Standard, as well as all other applicable regulatory standards.

3. Training: All applicable employees are required to be trained before starting work with hazardous materials or in hazardous conditions. Any employee that works with or around hazardous materials requires training:
   - To know about the hazard communication standard and what the standard requires
   - To know what material safety data sheets (MSDS) are, to understand how to read them and to know where they are kept
   - To understand all information on MSDS for each specific hazardous material that they may work with
   - To be able to recognize and understand hazard warnings, labels, and markings
   - To understand hazardous properties related to substances that they may use
   - To understand protective measures, equipment requirements, and personal protection requirements for working with each specific hazardous material
   - To be able to recognize exposure to dangerous substances through modes of exposure such as inhalation; absorption through the skin; ingestion, eating or swallowing; and through injection into the body circulatory system, or by cuts, abrasions and open wounds
   - To understand that exposures to dangerous substances can manifest symptoms in short amounts of time, acutely, and over long periods of time, chronically
   - To know what to do in case of emergencies and what methods of first aid are required when accidental exposures to hazardous substances occur
   - To know that hazard communication training is an on-going program. It is required anytime materials, methods, or procedures are changed.

4. MSDS or Material Safety Data Sheets are to be provided by manufacturers of all hazardous materials that are used in a given work place. MSDS are required to:
   - Be on site before hazardous materials are used.
   - Be routinely updated and replaced whenever chemical properties are changed by the manufacturer.
   - Be kept at locations that are readily accessible to all employees.

- Be kept on file by users for at least 30 years after hazardous products are no longer used.
- Exactly match inventories of hazardous materials being used on site, or that have been used on site.

5. Labels and Markings: All hazardous materials containers require marking by the manufacturer and hazardous materials containers shall be relabeled when the substance is transferred out of the original container. Hazard communication labels shall contain at minimum, the following information:

- Name of the substance
- Specific hazard(s) of the substance
- User protection requirements
- Target organs that may be affected by the substance
- Manufacturer name and other specific data

COMMUNITY RIGHT TO KNOW

Whenever community emergency personnel respond to stabilize emergency conditions at a given site, they and the surrounding community have the right to know significant inventories of hazardous materials that they may be subjected to while on site. The United States Environmental Protection Agency has prepared a document called the "EPA List of Lists" that shows quantities of substances that must be reported annually to the state emergency response commission, local emergency planning committee, and to the local fire departments that would respond to a given site. Please remember that each state and community has the right to be more stringent in regulating than the federal agencies may be. In that regard, someone at each site should research the full extent of regulatory liability, beginning with federal requirements, state requirements, and ending with local requirements.

The following emergencies can involve chemical storage inventories:

- Terrorist threats
- Fire and explosion
- Natural disasters such as storms, floods, earthquakes, power outages, and others
- Hazardous substance spills and releases, on site and to the environment
- Transportation emergencies
- On-site process, personnel, and equipment emergencies
- Off-site community issues, acts of violence and others

What Are Emergency Planning and Community Right to Know Issues

Following are the methods that emergency planners and the community use for informational purposes:

- Notifying local emergency response agencies about types of hazardous materials and conditions that may present significant hazards in times of emergency
- Annual Tier II Reporting of hazardous materials inventories stored on site. This involves mandatory and optional reporting quantities of dangerous substances

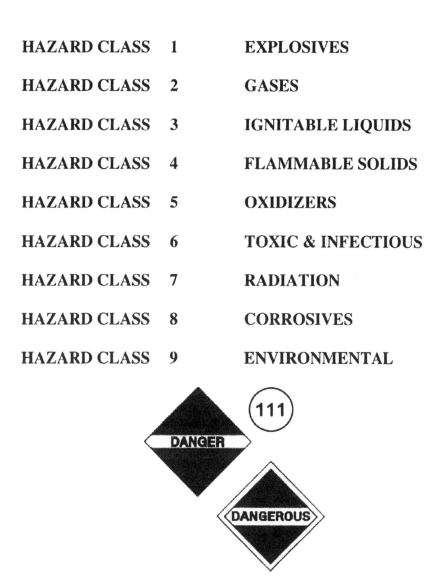

HAZARD CLASS	1	EXPLOSIVES
HAZARD CLASS	2	GASES
HAZARD CLASS	3	IGNITABLE LIQUIDS
HAZARD CLASS	4	FLAMMABLE SOLIDS
HAZARD CLASS	5	OXIDIZERS
HAZARD CLASS	6	TOXIC & INFECTIOUS
HAZARD CLASS	7	RADIATION
HAZARD CLASS	8	CORROSIVES
HAZARD CLASS	9	ENVIRONMENTAL

Figure 2.6 Department of Transportation hazard classes (DOT/ICAO, 49 CFR).

stored on properties, fixed equipment, and mobile equipment. Reports are sent to state and local emergency agencies.

- Annual Reporting of regulated materials usage, EPA Section 313, Form R. This involves annual reporting of quantities of regulated materials that were used to manufacture products, to process materials to make products, or otherwise used in methods or procedures conducted at a given site.
- Posting Placards: Several states and cities have established ordinances that requires posting of NFPA 704 Placards. Another means of hazard recognition, the NFPA placard serves to warn responding public emergency action personnel that chemical hazard potentials exist on a property or within a building. Unfortunately, property owners may not understand how to properly display or code the placard. Anyone that is responsible to comply with city ordinances requiring posting of NFPA 704 placards and needs direction in meeting compliance, should contact their local fire department or the National Fire Protection Association.

RECOGNITION OF CHEMICAL HAZARDS IN SHIPPING AND HANDLING

The United States Department of Transportation (DOT), International Civil Aviation Organization (ICAO) and State(s) Departments of Transportation, requires that all containers of shipped dangerous goods be labeled. The above organizations have worked with the United Nations to establish international standards for containers and markings.

The Occupational Safety and Health Administration (OSHA) requires that the DOT labels and markings remain on original containers, and that DOT labels on containers become part of site specific hazard communication and hazard recognition systems.

The International and U.S. DOT container labeling systems are composed of 9 hazard classes of chemicals as found in the North American Emergency Response Guidebook.

HAZARD RECOGNITION SUMMARY

Hazardous conditions may be related to any of the following:

- Chemical hazards
- Biological hazards
- Physical and site hazards
- Improper personal protection
- Lack of oxygen

Basic Criteria for Using the NFPA 704 Placard

All symbols listed below related to hazard potentials that may exist during emergency conditions:

ZONE COLOR CODES

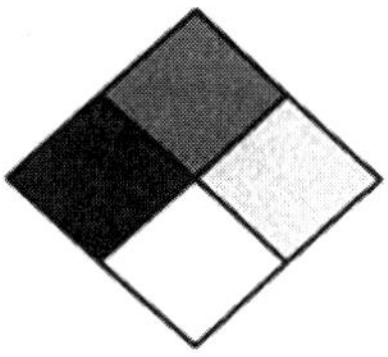

➢ RED   Fire Potential

➢ BLUE   Health Hazard Potential

➢ YELLOW   Reactivity Potential

➢ WHITE   Explanation of Specific Hazards Found On Site

NUMBER CODE

❖ 0   No Chemical Hazard in This Zone

❖ 1   Minimal Hazard Potential in this Zone

❖ 2   Moderate Hazard Potential in this Zone

❖ 3   Serious Hazard Potential in this Zone

❖ 4   Extreme Hazard Potential in this Zone

SPECIFIC HAZARDS

o OXY   Oxidizer Substances On Site
o W   Water Reactive Substances On Site
o COR   Corrosive Substances On Site
o ACID   Acid Hazards On Site
o ALK   Alkali Hazards On Site

o   Radioactive Hazards On Site

Figure 2.7 National Fire Protection Association.

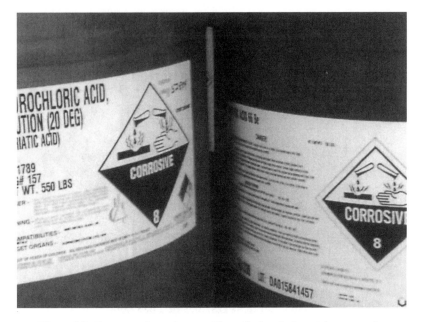

Figure 2.8 For proper hazard recognition, all chemical containers must be appropriately labeled (OSHA HAZCOMM 29 CFR 1910.1200).

- Too much oxygen
- Ionizing radiation
- Non-ionizing radiation
- Space configuration
- Site configuration
- Site layout
- Environmental hazards
- Equipment operation hazards
- Improperly maintained equipment
- Improperly maintained worksite
- Defective materials
- Structural integrity
- Fall protection
- Walking and work surfaces, slips, trips, and falls.
- Overhead activities
- Improper lifting and improper ergonomic practices
- Electrical hazards

- Noise hazards
- Pneumatic hazards
- Fluid pressure hazards
- Extremes of temperature
- Acts of terrorism and sabotage
- Non compliance of regulatory standards
- Improper standard operating procedures
- Improper management practices
- Improper work practices
- Lack of discipline and training in personnel

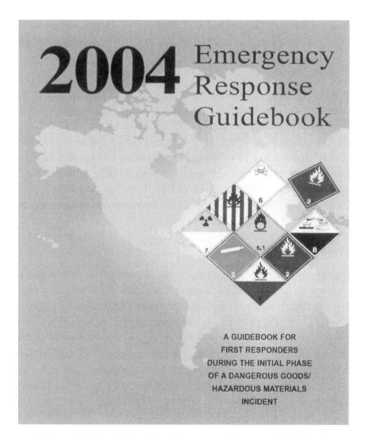

Figure 2.9 North American Emergency Response Guidebook.

HAZARD RECOGNITION DISCUSSION

Thus far, we have discussed situations and considerations that characterize issues related to evaluating hazardous conditions. We have already discussed some issues that are related to chemical and biological hazards as well as dangerous goods.

Some topics listed in this chapter have been already mentioned. However, we should ponder the following issues and consider OSHA guidance for each.

Physical and Site Hazards

Every work site varies; however each site contains hazards that are unique to it. There are times when obscure hazards need to be sought out and enumerated. For example, as four agents conducted an environmental property assessment, they crossed a beautiful grassy field. One of them stepped into a pothole and broke his leg. Upon helping the injured person from the hole, another agent stepped upon a sharp nail that punctured his shoe and was driven into his foot. As the two injured persons were helped from the field, one of the remaining agents tripped on a strand of barbed wire and fell onto a broken glass bottle. Got the picture? Physical site hazards happen.

Personal Health Status

Before anyone wears chemical protective clothing, especially a respirator, they need to be sure that they don't have a health problem that will be exacerbated by donning gear. OSHA requires passing a medical evaluation before a person is authorized to wear respiratory protection. When plastic, rubber, or polymer chemical protective clothing is worn, the wearer is subjected to extremes of heat that could cause heat stress and eventually heat stroke. People in poor health should not be placed in harm's way. Anyone that is taking mind or body altering drugs, irregardless if they are prescribed or illicit, should not be working in hazardous environments.

Improper Personal Protection

Using protective equipment that does not actually protect the wearer from suspected hazards is as serious as not wearing personal protection when it is needed. Lack of proper training, misinformation, and poor management is often at fault in this situation. Often we see workers performing mold and biohazard abatement work wearing paper dust masks and thin plastic, paper mesh coveralls over their work clothes. The workers believe that they are protected from legionella and other water borne pathogens; or they believe to be protected from chemicals, hepatitis, typhoid, or cholera contamination as they wade through sewer drainage in an effort to prevent the spread of mold spores in a water-damaged building. They are not protected, especially the ones wearing beards.

Lack of Oxygen

Also called anoxia, lack of oxygen is often a cause of death for persons that enter confined spaces. Normal air contains approximately 20.8% oxygen, 78% nitrogen, .93% argon, .03% carbon dioxide, and some other stuff. Safe operating oxygen levels range from 19.5 % to 21%. When oxygen drops below 16%, the environment becomes immediately dangerous to anyone not receiving supplied air. Anyone foolish enough to wear air purified respiratory protection at oxygen levels less than 19.5%, are toying with death.

Too Much Oxygen

Is there such a thing as too much oxygen? Oxygen is good, especially for lighting fires. No, oxygen does not burn; however oxygen makes everything else burn faster. Oxygen levels above normal greatly enhance burning.

Space Configuration

This may be a problem in your garage or closet. However, the person that enters a grain elevator to loosen a jammed mixing blade quickly learns that conical-shaped containers are hard to quickly move around in, especially if the mixing blade was not safety locked and the power to run the blade was not properly locked and tagged out. Many people become trapped in irregularly configured spaces every year.

Site Configuration and Layout

Who would build a parking lot or a camp ground at the bottom of a dam in a reservoir? When a chemical fire rages and emits toxic fumes into the community, emergency responders shall establish incident command centers up wind from the incident. Why? Don't move down wind of a fire. It may be the last thing you ever do. Some buildings may have only one way to enter and the same egress point to exit so what happens when fire or disaster closes that exit point? Site configuration.

Environmental Hazards

Environmental hygiene personnel are often challenged to determine why some employees complain that they become ill from the moment they arrive at work until they go home. By the way, sometimes this problem disappears after retirement; however actual environmental health issues do exist in most areas. Workers that enter unventilated confined spaces are often subject to dangerous environmental hazards. The National Institute of Occupational Safety and Health (NIOSH), Centers for Disease Control, EPA, and OSHA focuses upon the following as environmental hazards:

1. Air pollution and respiratory health in areas considered normally safe

2. Cause of health concerns such as asthma
3. Cancer potentials found in the air that we breathe
4. Airborne hazards in confined spaces and other poorly ventilated locations
5. Hazards yielded by extremes of weather conditions
6. Pollutions found in water, soils, and foods from normal biological activities as well as man-induced pollutions
7. Microbial hazards generated during water damage and increased humidity such as fungi (molds), waterborne bacteria (legionella), and others
8. Hazardous gases liberated from fire and combustion
9. Radiation from sources such as radon gas.

Equipment Operation

To sum up hazard potentials and causes of actual personnel injuries, it must be emphasized that it always helps to follow operating instructions exactly as they are presented by manufacturers of equipment. Lack of maintenance, operator error, and lack of common sense also are major findings in accident investigations.

Improperly Maintained Worksites

A clean and organized worksite will greatly reduce hazard potentials as compared to the alternatives. One man once said, "I've been mugged in nicer places than the place that I work." Cluttered walkways, blocked exit doors, and blocked utility access points are examples of hazardous conditions that do not need to exist. Equipment falling apart can be due to lack of commitment towards routine maintenance activities. The perspective of "pay me now or pay me later" applies to equipment and facility maintenance issues. Periodic maintenance programs will save any organization money as well as reduce potentials leading to health, safety, and environmental liabilities.

Defective Materials and Structural Integrity

During an OSHA inspection it is often noted that walkways, stairs, ladders, and handrails are not sturdy enough to safely use. After a flood, fire, or natural disaster, degradation of wood products from mold or wood rot can occur. All structures must be evaluated for safe use or habitation. Far too often in chemical operations it is found that installed equipment and utilities are manufactured from materials that are attacked by surrounding chemicals. For instance, stainless steel comes in several varieties and some do not resist strong corrosive chemicals. Plastics and other polymers can be attacked by varieties of organic solvents. We must be aware of resistance properties and reactive properties of all materials used in the workplace.

Walking and Working Surfaces—Slip, Trip, and Fall Prevention

During safety and health inspections, attention should be focused upon areas that are often overlooked. Our daily routines should contain workplace evaluation and, common sense must prevail. Even though a worker does not fear working in high places, he is required to be provided with fall protection. Refer to the OSHA database on the Internet for examples and regulatory guidance in this regard.

Overhead Activities

Overhead cranes and work activities provide additional hazards that need to be evaluated. Do all workers require head protection? In this case, yes. Does one type of hardhat meet all needs? No. Again, the OSHA database can provide guidance in overhead work considerations for both industry and construction.

Improper Lifting and Ergonomics Practices

In past years the Occupational Safety and Health Administration attempted to implement standards that would regulate ergonomics issues in the workplace. The standard was reduced to a voluntary standard, even though millions of dollars are spent each year by employers to treat employees with job induced musculoskeletal disorders, including back injuries. Again, these are hazards that normally exist. All employers should review the OSHA database for ergonomics guidance directives. The cost of an active ergonomics program is much less than employee treatment and lost time on the job.

Electrical Hazards

How many times do we find work activities in progress that have extension cords on the floor bathing in a puddle of water? Unfortunately, much too often. It's extremely frustrating when the site safety officer is told by the shift supervisor that, "It's OK. We have ground fault protection." Work activities that involve electricity require considerations such as specialized training, proven standard operating procedures, personal protective equipment, and OSHA control of hazardous energy or lock out/tag out training for affected employees.

Ventilation and Noise Hazards

OSHA regulates ventilation and occupational noise exposure under standards relating to occupational health and environmental control. Sometimes we are required to consider respiratory protection potentials for a variety of tasks and operations. The best choice for respiratory protection would be engineering controls and, in this respect, ventilation is the best alternative. Please know that everyone can't wear a respirator. However, everyone can inhabit a work area that is properly ventilated. Noise is considered to be pollution and it's not normal to spend your life with ringing ears. Strict guidance in establishing criteria for working in noisy areas is provided by OSHA. Please review the OSHA database for establishing hearing conservation programs in noisy work areas.

Pneumatic and Fluid Pressure Hazards

Should pressurized pipes and tanks be marked or labeled hazardous? It would be a great idea. In some cases warning labels are required, depending upon materials contained in pressure lines and potentials of hazardous pressure. Again, check the OSHA database.

Temperature Extremes

As we consider personal protection, temperature extremes become an issue. Anyone who has worn total coverage chemical protective equipment knows that heat exhaustion and heat stroke are only minutes away when working inside PPE or in high temperatures. Anyone that has suffered from frostbite understands the value of protecting exposed body parts from the cold.

Acts of Terrorism and Sabotage

Since the events of September 11, 2001, issues have emerged that have added more hazards to evaluate in the workplace as well as any place that people may reside. During 2003, the Department of Homeland Security, Federal Emergency Management Agency, and the Department of Transportation prepared and distributed training materials to:

1. Help workers become aware of terrorist potentials and to plan to deal with actual incidents, should they arise
2. Be able to work with public emergency response personnel, hospitals, and public heath agencies

Training materials, written and digital, are available at no charge from the U.S Department of Transportation (D.O.T.) and the Federal Emergency Management Agency (FEMA).

Noncompliance with Regulatory Standards

Every workplace should be periodically audited to determine current regulatory status. Please know that it is the responsibility of the owner or operator of every organization to determine their regulatory applicability and liability. In other words, the government does not send reminders to employers that inform them of what regulations cover them. How should this be handled? Hire an environmental, health and safety professional or surf the Net for hours to find applicable regulatory information. Some employers don't worry and wait to be inspected. Caution: some fines begin at $25,000 per day of non-compliance, or worse yet, employees are injured or killed because they were not properly protected. Sounds like potential felony allegations? Yes, they are.

Improper Management Practices, Improper Work Practices, and Improper Standard Operating Procedures

The employer is responsible to ensure that employees work in a safe and environmentally friendly work place. Employees are responsible to follow proper procedures and perform safe work practices. Sound good? Let's come back to the real world. How do organizations become out of compliance, if they ever were in compliance with all environmental, health, and safety regulations? Basic EHS regulatory requirements bore and intimidate people that are not committed to understanding them. Sometimes we get so busy to meet business objectives, we overlook regulatory compliance, safe work practices, and common sense. Again, how do we keep out of trouble? Hire an EHS professional or add regulatory compliance responsibilities to some job description. Engineers and managers beware; you are responsible to ensure that standard operating procedures are written to specifications that can be correctly and safely followed when an employee performs a task. Don't write them and proof them by yourself. If you do, your employees may take them as you have written them and in a troubled tone ask, "Are you sure you want me to follow this procedure just like it is written?"

Lack of Discipline and Training

Functional and written standard operating procedures must also meet all aspects of environmental, health, and safety compliance. The regulatory agencies view this as an act of due diligence. In other words, if everything is written correctly and regulatory compliance programs are functional and being followed by trained employees, the agencies will usually take a less harsh stance when they find something wrong. One item that is often overlooked is a human relations problem. What does an employer do when employees willfully neglect to follow procedures that lead to breaking the law? In this regard, an employer must take action or suffer the consequences. Human relations standards should be functional and written as well.

ABOUT TRAINING

Any training that is required by the many regulatory standards that cover most employers requires proof. What kinds of proof? Copies must be kept of curriculum provided. Persons providing training are required prove that they are qualified to teach. Proof of demonstrated competency for each employee trained is also required. For more information in this regard, refer to guidelines for training in each regulatory guideline. Another useful guideline is the ANSI Z490 Standard, Accepted Practices in Safety, Health, and Environmental Training.

HAZARD ASSESSMENT SUMMARY

Hazardous conditions and their associated risks must be evaluated before we subject ourselves and others to undefined situations. How we do whatever we plan to do, our protection, decisions that we make for proper operating procedures, and safe results of our endeavors can be influenced by topics listed below. Review this list, we will see these and similar topics again; how we deal with these conditions could change our lives.

☐ Chemical hazards
☐ Biological hazards
☐ Physical and site hazards
☐ Personal health status
☐ Improper personal protection
☐ Lack of oxygen
☐ Too much oxygen
☐ Ionizing radiation
☐ Non-ionizing radiation
☐ Space configuration
☐ Site configuration
☐ Site layout
☐ Environmental hazards
☐ Equipment operation hazards
☐ Improperly maintained equipment
☐ Improperly maintained worksite
☐ Defective materials
☐ Structural integrity
☐ Fall protection
☐ Walking and work surfaces, slips, trips, and falls
☐ Overhead activities
☐ Improper lifting and improper ergonomic practices
☐ Electrical hazards
☐ Ventilation and noise hazards
☐ Pneumatic hazards
☐ Fluid pressure hazards
☐ Extremes of temperature
☐ Acts of terrorism and sabotage
☐ Noncompliance of regulatory standards
☐ Improper standard operating procedures
☐ Improper management practices
☐ Improper work practices
☐ Lack of discipline and training in personnel

Figure 2.10 Hazard assessment summary.

PERSONAL PROTECTION

As previously mentioned, use of personal protective equipment should not be substituted for engineering controls, proper work practices, or administrative controls. Personal protective equipment is used in conjunction with these controls to ensure worker safety. Authorized use of personal protective equipment requires the user to have knowledge in hazards awareness and proper training for using personal protective equipment. Before using, personal protective equipment must be fitted to the user, and it must be clean and serviceable. Personal protective equipment must be selected to meet the job; both employers and employees are required to understand the purpose of the equipment and its limitations.

PPE Hazard Assessment

Employers are responsible for assessing the workplace for hazards that require using personal protective equipment. All hazards that require personal protective equipment to safely work shall be listed and certified in writing by the employer or the employer's representative.

PPE Training

Employees assigned to wear personal protective equipment must also be listed and trained. Documents shall be kept showing the hazards assessed, employee(s) trained to use specified personal protective equipment, and the types of training provided. Employers are responsible for certifying that employees have been trained to use personal protective equipment.

Employees' PPE training requirements include the following:

- Employees must be trained before using personal protective equipment.
- Users must know when and how to use PPE.
- They must understand equipment limitations, proper care, maintenance, useful life, and, in some cases, when to dispose of the equipment.
- Employees must know if more than one type of PPE will provide adequate protection and, in that case, they should be provided a choice.

Head Protection

Head injuries are often caused by falling or flying objects and by bumping into fixed objects. Usually provided as hats, head protection must be able to accomplish two objectives:

- Resist penetration
- Absorb shock or blow forces

Selection of Head Protection

Head protection is classed in the following types:

1. Type 1, which includes helmets having a full brim that is not less than 1 ¼ inches wide
2. Type 2, which includes helmets without a brim and having a peak that extends forward from the crown of the hat.

Industrial head protection involves 3 classifications:

- Class A - For general service and limited electrical protection
- Class B - For utility service and high electrical voltage protection
- Class C - For special service and contains no electrical protection
- For fire service personnel, head protection must contain ear covers and a chin strap along with other requirements that are listed in the OSHA regulation 29 CFR 1910.156 (e)(5).

Materials used to manufacture head protection should be water resistant and slow burning. Headgear must have a ventilation space between the headband and shell. It should be marked listing the manufacturer, ANSI designation, and classification. Helmets are date stamped by the manufacturer and should be replaced as designated by the manufacturer (usually 5 years).

Eye and Face Protection

Proper eye protection is required when potentials exist for eye or face injury such as flying particles, hot materials, chemicals, gases, vapors, and light radiation. Protective equipment is required to meet the following criteria:

- Provide protection against hazards they were designed to protect
- Be comfortable to wear
- Fit snugly and be durable
- Be easy to clean and capable of decontamination
- Kept clean and functional
- Must be marked with manufacturer's identification information
- Light radiation filter lenses must be marked to indicate shade number in order to facilitate choices of protection for specific operations.

Note: OSHA recommends that emergency eyewashes and first aid instructions be placed in all locations that contain eye hazards.

Hearing Protection

High noise levels will cause hearing impairment and hearing loss. It must be noted that people subjected to high noise levels undergo physical and psychological stress. Hearing damage is irreversible. Examples of hearing protection include:

- Pre-formed or molded earplugs
- Self-forming ear plugs
- Sealed earmuffs that can be adjusted to create a seal around each ear

CHEMICAL PROTECTIVE CLOTHING (CPC)

CPC is personal protective equipment that is designed to protect the wearer from chemical hazards. Chemical protective clothing should shield the wearer from dangerous substances by resisting any effect that a given substance may present. Selection and use of CPC should be based upon knowing the limitations of each specific grade of equipment. It must be pointed out that there is not any single combination of CPC that is capable of protecting against all chemical hazards. In this regard, assessment of chemical hazards and knowing CPC resistance towards chemical properties of dangerous substances must be part of selecting proper chemical protective clothing.

Employers are responsible to develop a site-specific chemical protective equipment (PPE) program. The following are key elements of a site-specific PPE Program:

- Identify all hazards that require use of personal protective equipment.
- Persons designated to use CPC must be medically fit to wear it.
- PPE should be selected based upon its protection factor towards substances that exist in a given location. Also, the equipment must be properly maintained and serviceable and properly fitted to the user.
- In using chemical protective equipment including respiratory protective equipment, it may be necessary to conduct environmental surveillance to guarantee that environmental concentrations of dangerous substances do not exceed PPE protection limits.
- Before using chemical protective equipment, it must be decided if contaminated equipment will be decontaminated or disposed.
- Before donning any type of personal protective equipment, the user must be trained to safely use the equipment.

PPE Selection

Chemical resistant personal protective equipment must be able to resist:

1. Permeation or soaking by dangerous substances.
2. Degradation or chemical breakdown by dangerous substances.
3. Penetration or eating through the PPE by dangerous substances.

Chemical protective PPE can be permeated if it is manufactured from products that will react with a given dangerous substance. Certain chemical factors can influence PPE permeation rates such as:

- Concentration of a given dangerous substance
- Environmental and physical factors such as humidity, pressure, and temperature

Effective use of chemical protective PPE depends upon several factors such as:

1. Durability of PPE construction materials—They must be able to resist tears, punctures, abrasion, and repeated use.
2. Flexibility of PPE materials—It will not interfere with completing tasks performed in a hazardous environment. This is a serious consideration when selecting proper gloves.
3. Temperature effects—especially in cold environments that often stiffens most PPE.
4. Ease of decontamination—in some cases it may be most feasible to use "dry decontamination procedures" or disposable PPE.
5. Equipment compatibility—in some cases, it may be difficult to wear head protection with total coverage PPE.
6. Duration of use—chemical resistant personal protective equipment must be able to withstand extended exposures to dangerous substances without protective materials breakthrough or degradation.

CONDITIONS REQUIRING SPECIAL PERSONAL PROTECTIVE EQUIPMENT

Fire response personnel are required to wear supplied air respiratory protection and flame and heat resistant clothing - often referred as "turn out gear." Firefighter turn out gear is not considered chemical protective clothing; in fact, people that respond to chemical releases wearing firefighting equipment are taking unwarranted chances with their well-being. Note: per HAZWOPER regulations, "firefighting teams are not HAZMAT teams."

Radiation protective equipment and methods and procedures relating to safe work activities are described in guidance provided by the U.S. Department of Energy and the Nuclear Regulatory Commission. Radiation hazards and the specifics relating to methods, procedures, and personal protection are discussed later in this text.

Confined space entry and rescue operations require personal protective equipment and chemical protective clothing as work sites may vary. In all confined space operations, entry personnel are required to wear retrieval harness and related equipment. A more comprehensive coverage of confined space work operations as well as equipment, applications and procedures will be presented later in the book.

LEVELS OF PROTECTION

Personal protective equipment is categorized in four primary groupings:

1. Level D, which is considered basic chemical protective work clothes, does not involve respiratory protection. Level D protection is authorized when atmospheric hazards have been evaluated and hazards will not affect the skin or the respiratory system.

2. Level C involves using total coverage chemical protective clothing and air purified respiratory protection. Typical Level C protection includes a plastic or rubber coverall suit with hood, rubber gloves, rubber boots with steel shank in the soles, and full face air purified respirator (cartridge mask). It is highly recommended that cotton underwear and a lightweight, Tyvek total coverage inner suit be used as first line of defense and a covering for the underwear. Inner gloves are also recommended, especially as a contingency for potential tears that may occur in outer gloves. Duct tape is used to seal all areas between suit and gloves; respirator and suit; and zippers and suit. Don't forget to fold over a tab on exposed duct tape, in order to have a grip for use in removing during decontamination and doffing of the Level C apparel. Use of Level C protection requires that concentration of air contaminants be known and that filter cartridges have the ability to remove all air contaminants. Air monitoring is required when using Level C protection.

3. Level B protection involves the same total coverage by chemical protective clothing as Level C; however supplied air respiratory protection is used. The self- contained breathing apparatus used in Level B operations should be positive pressure self-contained breathing apparatus. Head protection can be worn when head injury hazards are present. In situations where airborne health hazard contaminant levels are unknown, the lowest level of protection should be Level B, and, in some cases, Level A or maximum protection should be worn.

4. Level A protection is considered maximum protection against chemical and biological hazards. Level A protection involves total encapsulation of Level B protection by a chemical protective, totally sealed outer suit. Level A suits contain extra strength gloves and foot encapsulation built into the suit. Personnel authorized to wear Level A protection must be trained to deal with potentials of having problems with positive pressure, self-contained breathing air and air hose supplied air systems. This involves disassembly of the air cir-

Figure 2.11 Chemical protective suits (Level A protection on the left and Level C protection on the right).

cuit in the suit and breathing remaining air within the suit while evacuating to a less hazardous environment *before opening the suit*.

Flash Protective Suits

In all levels of protection listed in this section, please know that anytime ignitable hazards exist, flash protection must be worn over personal protective equipment. Flash protection usually involves a total encapsulation suit, minimum level B protection that resists the initial flash of fire. Flash protection only protects the wearer from fire flash and not heat of burning. When fire flash occurs, entry personnel must immediately evacuate to safe locations. An initial flash of fire lasting less than one second will cause non-flash protected personnel to suffer second and third degree burns. At no time should personnel enter an atmosphere greater than 10% of the lower explosion limit of any given ignitable substance.

RESPIRATORY PROTECTION

The fastest way that contaminants can enter the body is through the respiratory system. Uses of respiratory protective equipment are serious, and please know that not everyone can wear a respirator. During 1998, the Occupational Safety and Health Administration implemented respiratory standards that placed stringent controls for using respiratory protection in all industries.

Respiratory Protection Program

Any employer that desires employees to wear respiratory protection is required to develop and implement a functional respiratory protection program. Key elements of a respiratory protection program are:

- A written plan that explains why respiratory protection is required above engineering controls and other safe work practices
- A program that describes how respiratory protection shall be selected, as compared to on-site potential respiratory hazards
- Mechanisms for medical evaluation of all respirator users must be described
- Mechanisms for fitting all respirator users with a respirator that comfortably fits each respective wearer, qualitatively and quantitatively
- Mechanisms for refitting each respirator user at least annually, unless otherwise directed in the most current OSHA Respiratory Protection Standard that may apply to specific industry applications. In some industry applications, respirator wearer must be refitted 2-3 times annually.
- Formal training is required for all respirator users and it shall include:
  1. Training on selection and proper use of assigned respirators
  2. How to inspect, clean and disinfect assigned respirators
  3. How to properly store assigned respirators
  4. How to handle problems with damaged and defective respirators

Kinds of Respirators

The most common types of respirators are:

- Paper dust masks: Respirators required for use in dust hazard applications.

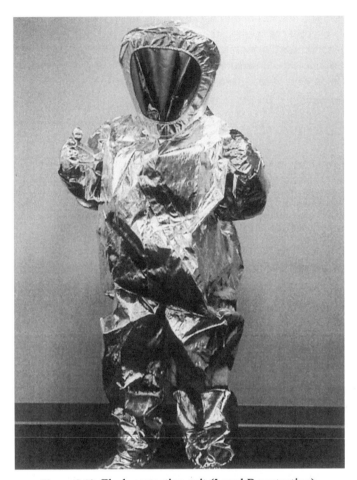

Figure 2.12 Flash protection suit (Level B protection).

- Air Purified Respirator (APR): To be used only in known hazardous environments, such that air concentrations of specific hazards are known and filter cartridges have the ability to filter out airborne hazards. Cartridges used in APRs are color-coded for specific hazards and some contain end of service life (ESLI) indicators built into the cartridge. APR users require training to know when cartridges are due for changing. As mentioned before, periodic air monitoring is likely to be required when wearing air purified respiratory protection in hazardous environments. APRs are often used as ½ masks in low hazard environments, and full face masks in higher hazard environments.

- Supplied Air: Self-contained breathing apparatus (SCBA) or air tanks worn by users along with mask and air distributing systems. Another supplied air respirator is air line-supplied breathing systems. Either can supply breathing air required for conditions that have potential for being immediately dangerous to life and health (IDLH). When using air line systems in an IDLH environment, an emergency escape SCBA system is usually required.

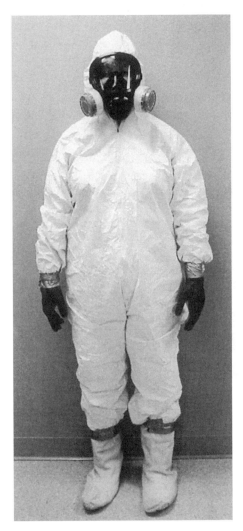

Figure 2.13 Level C chemical protective suits require air purified respiratory protection.

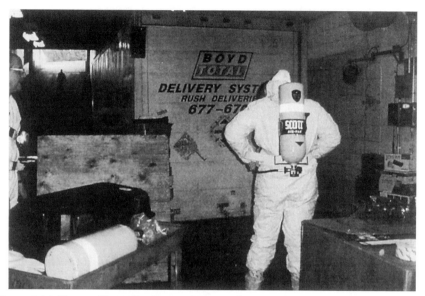

Figure 2.14 Hazmat Team Response often requires Level B chemical protection that requires positive pressure SCBA respiratory protection.

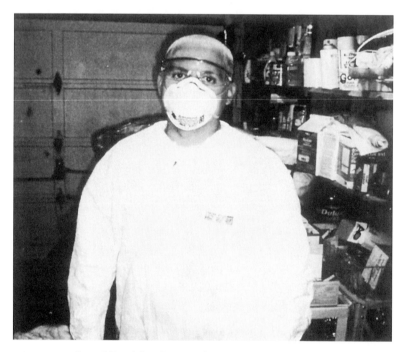

Figure 2.15 Level D minimal protection using a dust mask for respiratory protection from nuisance dusts.

STANDARD OPERATING PROCEDURES

SAFETY AND HEALTH MANAGEMENT PROGRAM

Your Company

Program: Safety and Health Management Program

Revision: Date

Supersedes: Other previous written procedures

Implementation Date:

A. PURPOSE AND SCOPE

All employees and visitors at YOUR COMPANY are protected by the Company Safety and Health Program. This program also applies to hazards covered by the General Duty Clause and other applicable OSHA, state, and city, safety and health safety standards

B. APPLICABILITY

All Company operations and worksites

C. REFERENCE

29 CFR 1900.1
29 CFR 1910 Subpart A-Z
ANSI Z490

D. GENERAL

YOUR COMPANY has developed this safety and health program for the purpose of managing safety and health activities that will reduce injuries, work related illness, and potential fatalities by taking pro-active approaches for safety and health regulatory and "common sense" compliance. This program applies to all company related activities, company properties, and respective company job sites.

This program has the following properties:

- (i) Management, leadership and employee participation
- (ii) Hazard identification and assessment
- (iii) Hazard prevention and control
- (iv) Information systems and training
- (v) Evaluation of program effectiveness

E. MANAGEMENT LEADERSHIP AND EMPLOYEE PARTICIPATION

YOUR COMPANY shall:

- Designate program responsibilities for managers, supervisors, and employees with the goal of ensuring the highest quality of safety and health in the workplace; and to hold them accountable for these responsibilities.
- Empower managers, supervisors, and employees with authority, to access relevant information, to conduct training, and to provide them with resources needed to execute their safety and health responsibilities.
- Designate at least one manager, supervisor, and or key employee that will receive and respond to reports relating to workplace safety and health conditions.

F. EMPLOYEE PARTICIPATION

YOUR COMPANY shall provide opportunities for employees to participate in developing, implementing and evaluating workplace safety and health programs. This would be done by using employee safety teams that:

- Routinely communicate with employees about workplace safety and health issues;
- Work with employees to ensure that they are informed about specific programs;
- Encourage co-workers to become aware of hazards that they make work with.
- Help employees become aware of their responsibility to report job-related incidents, injuries, illnesses, and hazardous conditions; and to stress working with management to promptly make recommendations about appropriate ways to control those hazards;
- Provide prompt responses to employee reports and their recommendations;

YOUR COMPANY shall encourage employees to make reports and recommendations about fatalities, injuries, illnesses, incidents, or hazards that may occur in the workplace, and shall encourage participation in the company safety and health program as directed by the National Labor Relations Act.

G. HAZARD IDENTIFICATION AND ASSESSMENT

YOUR COMPANY shall identify and assess hazards that employees may be exposed to, and evaluate work place safety by:

- Conducting self inspections ;
- Reviewing established safety and health programs;
- Evaluating hazards associated with new equipment, materials, and processes before using them in the workplace;
- Identifying hazards and immediately correcting uncontrolled dangerous conditions found on YOUR COMPANY property.

YOUR COMPANY will conduct hazard identification, inspection, and assessment procedures at least:

(i) Initially when changes are made in facilities, materials, methods and or processes;

(ii) As often as necessary to ensure compliance with all regulatory standards and use of best business management practices, a minimum interval of at least annually;

(iii) Whenever new safety and health information is received, or when changes occur in workplace or job site conditions that indicates that hazards are potential;

YOUR COMPANY will investigate each work-related injury or illness, or incident including near-hit incidents that have potential to cause injury to personnel, or damage to equipment or properties. This information will be recorded and retained indefinitely.

H. HAZARD PREVENTION AND CONTROL

YOUR COMPANY focuses upon hazard prevention and control, in accordance with all applicable regulatory standards.

YOUR COMPANY develops and maintains written and functional plans for regulatory compliance, and performs hazards assessment, self inspection, and correction of uncontrolled conditions.

I. INFORMATION AND TRAINING

YOUR COMPANY shall make sure that:

- Employees are provided information and training that is appropriate to company programs and that complies with regulatory directives.
- Employees are provided with appropriate information and training that will help them to perform work activities in a safe and environmentally friendly work place.

YOUR COMPANY shall provide information and training that includes:

- Hazards that employees may be exposed to and how to recognize them;
- Hazard control methods;
- Protective measures to prevent, or minimize employee exposure to potential hazards;
- Understanding applicable regulatory standards.

The following information and training shall be provided as follows:

- For personnel currently employed by YOUR COMPANY that have documentation of previous training and the employee can demonstrate knowledge and skills acquired from previous training, will not need to retake training unless otherwise directed by regulatory standards. All personnel must receive refresher training at least annually.
- To all new employees, before beginning work in or around hazardous conditions.
- As often as necessary to guarantee that employees are adequately informed and trained.
- Whenever a change in workplace or job site conditions indicates that a new or increased hazard or hazardous condition exists.

YOUR COMPANY Management will ensure that all employees receive information and training necessary for them to perform job assignments and will support their efforts in properly accomplishing their work assignments in a safe and regulatory compliant atmosphere.

J. EVALUATION OF PROGRAM EFFECTIVENESS

YOUR COMPANY will evaluate all company safety and health programs to ensure their effectiveness and applicability to workplace conditions. This evaluation will occur as often as necessary and at least annually. If any evaluation indicates deficiencies the program shall be revised in a timely manner.

K. MULTI-EMPLOYER WORKPLACE OR WORKSITE

Whenever YOUR COMPANY is considered to be a host employer, the following shall occur:

- Information regarding hazards, controls, safety and health rules, and emergency procedures shall be provided to all employers at the workplace or job site.
- Safety and health responsibilities shall be assigned as appropriate to other employers at the workplace or job site.

Contract employers that may work on YOUR COMPANY properties are responsible to:

- Ensure that YOUR COMPANY employees are aware of the hazards associated with contract employer's work and specific operating procedures that will be performed by the contract employer's personnel. All contract employers shall be responsible to coordinate all work activities with YOUR COMPANY EHS Officers or employees assigned to represent company EHS officer responsibilities.
- YOUR COMPANY shall be advised of any hazards or hazardous conditions that the contract employer identifies while performing at the workplace or job site.

L. RESPONSIBILITIES

The following personnel are responsible for managing YOUR COMPANY Safety and Health Program, as prescribed by federal, state and city regulatory guidelines:

1. Environmental, Health & Safety (EHS) Officer:
   OfficeTelephone:
   Cell Phone:
   Signed_____ EHS Officer
2. AssistantEHSOfficer:
   Office Telephone:
   Cell Phone:
   Signed_____ Asst. EHS Officer
3. Facility Manager:
   Office Telephone:
   Cell Phone:
   Signed_____ Facility Manager

Chapter 3

DANGEROUS SUBSTANCES

PROPERTIES OF DANGEROUS SUBSTANCES

As we begin our study of materials that can become dangerous substances, please be forewarned that dangerous conditions can result from materials that are not considered obvious hazards. Some substances become dangerous when secondary materials become intimately involved with them. For example, linseed oil is used in a number of "safe" products and is considered to be the product of choice for many oil base products. In past years the printing industry was charmed by new linseed oil containing "non-hazardous" cleaners that were not regulated as hazardous wastes or considered dangerous substances. Unfortunately, many printers learned that rags saturated with the oily cleaner on Friday initiated pressroom fires by Sunday. By the way, material safety data sheets (MSDS) should point out reactive hazard potentials.

To begin this section, let's build upon key definitions that should be considered by every user of potentially dangerous substances. Please know that the following definitions are basic and are presented from common sense perspectives. Unfortunately some regulatory agencies and professional organizations have produced conflicting "HAZMAT" definitions in order to satisfy their individual agendas. The author of this publication would rather provide the reader with easy-to-understand information that is to the point and user friendly:

- HAZARDOUS MATERIAL: Any material that can potentially cause harm or present dangerous conditions capable of causing negative effects upon people, animals, plants, and/or the environment.
- HAZARDOUS WASTE: Hazardous materials that are regulated by federal, state, or local agencies that no longer have value.
- IDLH: Levels or amounts of dangerous substances that if a person becomes over-exposed, will cause serious health conditions including death.
- VAPOR DENSITY: A physical comparison between weights of airborne substances and air. This relates primarily to those things that will blend into the air. When a liquid dangerous substance is spilled and it rapidly dries, where does it go? It usually becomes airborne as a vapor. In this business we want to know where dangerous substances go as they enter into the environment. The average molecular weight of air is 29. The average molecular weight of a substance is determined by adding atomic weights of each element that is contained in a molecule and denotes properties of the substance. The average molecular weight (MW) of gasoline is approximately 72 and compared to air, gasoline is 2 ½ times heavier than air (72/29 = 2.48). Gasoline vapors will normally stay close to the ground. Please note that most materials will not rapidly rise thousands of feet to destroy the ozone layer as some may suggest, due to that strange force that we call gravity.
- VAPOR PRESSURE: Molecules of substances that become airborne (volatile) bounce around inside of the container that stores them. The energies of molecular bounces are vapor pressure. As temperatures increase, so do the number of molecular bounces and the pressures against container walls. Whenever contained liquids (or gases) are exposed to fire or increased temperatures, an explosion is a potential hazard. Please remember the term BLEVE: Boiling Liquid Expanding Vapor Explosion.

MATERIAL FORMS

Dangerous substances exist in all material forms. How we process, handle, or implement standard operating procedures depends upon physical forms of each respective substance:

- Aerosols: Liquids or solids that are suspended in the air
- Bioaerosols: Microbiological materials that are suspended in the air, including spores, biotoxins, microbes, vapors, or droplets containing microbial agents or bi-products
- Dusts: Small pieces of dry substances that can be carried throughout an environment by moving air. Dusts eventually settle onto all surfaces, including operating computers and television screens. All dry substances including microbial spores, insects, and insect parts are considered constituents of dusts.

- Fumes: For the purposes of this text, fumes are vaporized solids that are often toxic and may or may not be visible to the eye may be called smoke, gases, vapors and other things. Welders beware; you are vaporizing metals that become toxic fumes. Firemen wear respiratory protection because danger lurks within fumes and gases generated by the flame.
- Gases: Substances that demonstrate fluidic properties. Gases have vapor pressures that exceed 40 (psia) pounds per square inch (absolute) at 100 degrees Fahrenheit.
- Liquids: A fluid having vapor pressures that are less than 40 (psia) pounds per square inch (absolute)
- Mists: Environmental vapors exposed to condensation temperatures
- Solids: Substances built by molecules that normally exist in a fixed position. Solids are substances that resist expansion, have definite shape, maintain volume, resist compression, are dense, and mix or diffuse into other substances very slowly.
- Vapors: As temperatures increase and the substance boils, it is emitted into the air as vapors. As temperatures cool, vapors condense and become solids.

CORROSIVE HAZARDS

Corrosive substances exist in a variety of material forms and are capable of destroying human, animal, and plant tissues. Corrosive substances etch and erode metals

Figure 3.1 DOT Hazard Class 8 - Emergency Response Guide 153.

and other materials. Acids are corrosive and so are alkalis. Acids are called acids, whereas alkali substances are often referred to as caustics and hydroxides. See Figure 3.2 below.

What About pH?

We can use a fancy definition, but who needs one? If comprehensive definitions are needed, look in the glossary. Dangerous substance workers should focus upon pH as being a numerical value that represents acidity or alkalinity of corrosive liquids and vapors. The symbol pH deals with properties of hydrogen. When we look at acids, hydrogen is shown in an acid molecule as H+ or a positive hydrogen ion. In alkali substances, hydrogen is shown connected to oxygen as a negative hydroxide ion or OH- . The best part of dealing with pH is that it can be measured numerically with digital reading meters or by using "old time" pH papers that turn colors in ranges often represented as shown in Figure 3.2. In all reality, concentrated acids at pH ranges less than 2 will likely dissolve a person's body and automobile.

Concentrated alkali substances will likely do the same thing at pH ranges that exceed 12. Acids and alkali are usually presented as percent or as specific gravity (compared to the density of water). For instance, 70% Nitric Acid is very dangerous stuff even though 30% of the mixture is water. Corrosive substance concentration is the prime determinant for hazard potential. pH measurement helps us to know if we are dealing with an acid or alkali.

Remember, pH is a measurement that indicates if a liquid or vapor is acid, alkali or otherwise. It is the concentration of the substance that promotes corrosive action. However, pH measurement is a quick and easy way to consider corrosive potential. EPA regulates corrosives as hazardous waste when acids have pH less than 2 and alkali have pH greater than 12.5.

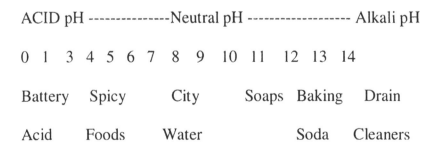

ACID pH --------------Neutral pH ------------------- Alkali pH

0 1   3 4 5 6 7   8   9   10 11   12 13 14

Battery   Spicy   City   Soaps Baking   Drain

Acid   Foods   Water   Soda   Cleaners

Figure 3.2 Continuum of corrosive substances.

Corrosive Hazards - Volatility and Reactivity

Some dangerous properties of corrosive substances include:

- They may readily disperse into the air (become volatile).
- Each substance has specific vapor densities (heavier or lighter than air).
- Acidic and alkali substances are capable of destroying living tissues.
- Acidic and alkali substances are capable of dissolving metals and other materials.
- Corrosive reactions may cause materials to ignite.
- Some corrosive reactions can induce explosion.
- Most corrosives are toxic health hazards.
- Some may take part in hazardous polymerization.
- Corrosives are water reactive (add corrosives to water, not water to corrosives).
- Some are oxidizers.
- Some corrosives may generate ignitable vapors and gases.
- Most corrosives react with many varieties of materials.
- Many are unstable, especially when left open to the air.

Workers should understand physical properties as volatility and vapor density for corrosive substances and other materials that they deal with. Also, personal protection, process control, and effective chemical handling require employees to be knowledgeable about process chemical hazards. All dangerous substance workers must be familiar with information provided in material safety data sheets (MSDS) for every hazardous substance that they may encounter.

Reducing Hazardous Properties of Corrosive Substances

To reduce corrosive hazards, we can neutralize a corrosive substance by properly reacting it with a "weak opposite." What are we talking about? The opposite of acids are alkalis and vise versa. What we shouldn't do is neutralize them with strong opposites. If strong acids are used to neutralize strong alkalis or if strong alkalis are used to neutralize strong acids, dangerous stuff can happen. For example, heat is generated (exothermic reaction) along with generating fumes, mists, and vapors. Some reactants can yield an explosion. Strong acids are usually neutralized with weak alkali substances such as sodium bicarbonate or soda ash. Strong alkalis are often neutralized using a weak acid such as vinegar, citric acid, and others. We will cover more about neutralization and reducing hazardous properties of dangerous substances later in this chapter.

Please note that neutralizing a corrosive substance may not eliminate all hazardous or regulated properties it may have. Some corrosive substances contain toxic properties or may have other hazardous considerations that cannot be changed by neutralization. Learn as much as possible about corrosives and other dangerous substances before working with them.

ELECTRONIC CHARGES AND ENERGIES OF ATTRACTION - THE GLUE THAT HOLDS EVERYTHING TOGETHER

You may already know this, but just in case you forgot; everything is assembled on this planet using a unique schematic. Remember that electronic charges (+) (-) and energies of attraction pretty well hold people, plants, air, rocks and everything else together. Where do the charges come from? From energies stored within the nucleus and sub-atomic particles that make up elements (or atoms). Atoms construct the stuff (or matter) that we deal with here on Earth, as well as on other planets within our solar system.

Acids

Substances containing one or more hydrogen ion (H+) and having the ability to liberate hydrogen gas when reacting with metals and other materials are acids. What do we mean by hydrogen ions? Let's see how acids are assembled from a molecular perspective. Remember: In physics, chemistry, and for some people, opposites are attractive to one another as shown in Figure 3.3.

About Acid pH

When liquid or gaseous acidic substances exist at pH levels less than 3, use extreme caution in handling. Also, the more concentrated (containing less water) acids are, extreme caution is needed in handling. The U.S. Environmental Protection Agency regulates any acid having pH 2 or less. In this regard, EPA regulates pH 2 or less waste materials as hazardous waste. Regulated acid wastes cannot be disposed without special treatment. Be careful, even if you neutralize a substance such as chromic acid, it still remains hazardous and regulated because chromium in the acid is toxic and a carcinogen. A number of other acids also fall into this category; it's important to know all properties of dangerous substances that we may deal with.

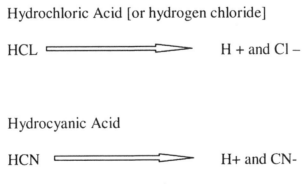

Hydrochloric Acid [or hydrogen chloride]

HCL $\Longrightarrow$ H + and Cl –

Hydrocyanic Acid

HCN $\Longrightarrow$ H+ and CN-

Figure 3.3 Opposites attract.

Alkali

Alkali chemicals are also called caustic, base, and hydroxide. Alkali chemicals contain hydrogen as (OH)- or hydroxide. As was shown in Figure 3.3, a similar balance of positive and negative charges exists between elements found in alkali chemicals. For instance, let's take a look at sodium hydroxide (also called caustic soda) and ammonium hydroxide (also called ammonia) shown in Figure 3.4.

About Alkali pH
When liquid or gaseous alkaline substances exist at pH levels greater than 12, use extreme caution in handling. Also the more concentrated (less water) alkali chemicals are, use extreme caution in handling. EPA regulates alkali, caustics, or bases at pH levels 12.5 or higher. The Environmental Protection Agency regulates waste alkali chemicals at pH 12.5 or higher as hazardous wastes. Regulated alkali chemicals cannot be disposed without special treatment. As mentioned about acids, some alkali chemicals may be neutralized but still remain regulated due to toxic or other hazards that remain after neutralization.

WHAT ABOUT NEUTRALIZATION?

We can decrease corrosive hazards in acids or alkalis by applying the reaction called neutralization. Neutralization occurs when an acid (or acid oxide) reacts with an alkali (or alkali oxide) and the end result is formation of a salt and water. The end result of neutralization is a treated product that has pH readings of pH6 to pH8. Let's take a look at combining sodium hydroxide and hydrochloric acid as shown in Figure 3.5.

What did it make? Table salt (NaCl) and water (H_2O) and lots of heat (exothermic reaction). Did you see that the opposite charges in each substance found one

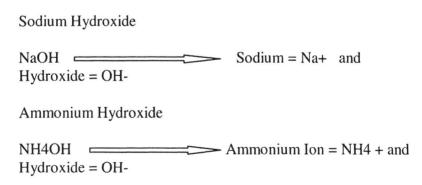

Figure 3.4 Sodium hydroxide and ammonium hydroxide.

Sodium Hydroxide (or caustic soda) (NaOH) = Na+ OH-

Hydrochloric Acid (HCL) (hydrogen chloride, muriatic acid and others) = H+ Cl-

Na+ OH-   +   H+ Cl- $\Longrightarrow$ Na+ Cl-   + H+ OH- + Temperature

Figure 3.5 Combining sodium hydroxide and hydrochloric acid.

another? When we combine concentrated, strongly opposite corrosive substances, we usually initiate a wild uncontrolled reaction. The extremes of pH and lack of water concentration in each substance usually generates heat (called exothermic reaction), vapors, and other unwanted physical properties. Should neutralization normally be performed as shown in Figure 3.5?

What if you were involved in cleaning up concentrated liquid caustic soda that was spilled onto the floor of a semi-truck? Danger: Don't neutralize as shown in Figure 3.5! Why? It's hazardous and may generate enough heat to cause a fire and generate airborne hazards. Plus, if you use too much acid to neutralize the alkali, the end result will be an acid spill. Always neutralize a weak corrosive with a weak opposite pH substance.

Neutralize Using a Weak Opposite

Turn back to Figure 3.1; wouldn't a weak opposite of a strong acid (pH < 1) be something that has a weak alkali pH? How about using baking soda or soda ash? Either is a good choice and, if you add too much, the end result can be easily adjusted to meet disposal standards.

Neutralizing Hydrochloric Acid (H+ Cl -) (pH ranges less than 2)
In another example, consider neutralizing Hydrochloric Acid and Caustic Soda using their weak opposites:

1. Using Baking Soda, or Sodium Bi-Carbonate ($NaHCO_3$)
   - Also called Sodium Hydrogen Carbonate, liberates lots of Carbon Dioxide
   - Good to use for small acid spills, applied as powdered soda
2. Using Soda Ash, or Sodium Carbonate ($Na_2 CO_3$)
   - Also liberates lots of Carbon Dioxide
   - Good to use for larger spills because it takes less to neutralize acids than sodium Bi-Carbonate

Neutralizing Caustic Soda (NaOH) (pH greater than 12.5)

1. Using vinegar, also called dilute (3-5%) acetic acid (CH_3 COOH)
2. Carbon dioxide, hydrogen gas, and heat are liberated as the solution becomes neutral.
3. Other mild acids are also used to neutralize alkalis; more will be covered on neutralization of corrosives later in the book.

The Environmental Protection Agency (EPA) and Department of Transportation (DOT) regulates corrosives as acids and alkalis. As previously mentioned EPA/RCRA regulate corrosives by pH. Even though concentration of a corrosive is most relative to its hazard potential, EPA uses pH as the hazardous waste indicator. Remember EPA and state regulatory authorities regulate acids as hazardous waste if they have a pH of 2 or less and alkalis as hazardous waste at pH 12.5 or higher. Caution: Cities and other local authorities may regulate corrosives at other pH ranges, especially as criteria for wastewater. Many cities in the United States require corrosives to be treated to the range of pH 6 to pH 8 before disposing in wastewaters, even though the pH of city water often exceeds pH 9. The Department of Transportation regulates corrosives based upon hazard potential. If a corrosive substance can damage animal or plant tissues or dissolve metal, it will most likely be regulated and require handling and marking as DOT Hazard Class 8 corrosive material.

IGNITABLE HAZARDS

Ignitable hazards will easily catch fire, rapidly burn and, in some situations, explode. They exist as many other things: solids, liquids, vapors, gases, dusts, and others. Properties of ignitable hazards include:

- Flash Point - Ignitable substances that become air will transform into airborne molecules at temperatures that are specific to each substance. Most gasoline mixtures start to become air at -42 degrees Fahrenheit (F). Toluene starts to become air at 30 degrees F. Releases of airborne molecules occur at liquid surfaces as vapors are released from containment. Temperatures at which materials emit ignitable airborne molecules are called flash points.

- Air Concentration - As an ignitable substance blends into the air and the concentration of the substance increases, fire danger potentials also increase. Each ignitable substance has a given air concentration where, when an ignition source is present, it will burn. This concentration point is called the lower explosion limit (LEL), also called lower flammability limit. Burning will continue within an air concentration range that is specific to each substance. When the air becomes too concentrated, burning will stop. This is called the upper explosion limit (UEL,) also called upper flammability limit. Anyone that has

Figure 3.6 DOT Hazard Class 3 - Emergency Response Guide 127.

been stranded in a car with a flooded carburetor has witnessed this effect. LEL, Burning Range, and UEL are illustrated in Figure 3.7.

- LEL: The lowest air concentration of an ignitable substance that will support ignition.
- UEL: Air concentration where too much ignitable substance is present to burn.

Note: LEL, Burning Range and UEL is usually listed as percent (%) or parts per million (ppm).

Percent to PPM

The comparison of percent (%) concentration to parts per million (PPM):1% = 10,000 parts per million. Multiply percent by 10,000 to determine parts per million. Note: the LEL for Toluene is 1.1%, or 11,000 parts per million (ppm).

Vapor Pressure

Have you ever noticed what happens when you remove the cover from the gas tank of your automobile? During warm weather it will "whoosh" as it whispers, "Feed me." The whooshing sound comes from pressure caused by warm gasoline mole-

CONCENTRATION EFFECTS

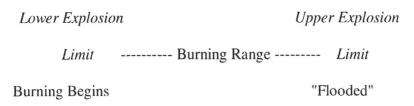

Lower Explosion   *Upper Explosion*

Limit   ---------- Burning Range --------- *Limit*

Burning Begins   "Flooded"

Figure 3.7 Explosive concentration effect.

cules bouncing around inside the tank and, when the cover is removed, varieties of gas molecules escape from the tank. This is called vapor pressure. All substances that become air have specific vapor pressures at given temperatures.

Vapor Density and Inadequate Ventilation
As already mentioned, knowing the molecular weight of a hazardous airborne substance and comparing it to the weight of air (29.1) helps us to predict locations it will congregate within an enclosed area. Ignitable atmospheres at LEL concentrations are dangerous fire or explosion hazards. In fact, entry should not be allowed in ignitable atmospheres greater than 10% of the LEL.

EPA regulates ignitable hazardous wastes; DOT regulates ignitable liquids, vapors and gas.

Ignitable wastes having flash points less than 140 degrees Fahrenheit are regulated by EPA as hazardous wastes. The Department of Transportation (DOT) regulates ignitable liquids at flash points less than 140 degrees Fahrenheit and requires labels and placards as flammable liquids, DOT Hazard Class 3. Of course, hazardous wastes are regulated by the Department of Transportation and are transported following DOT regulatory parameters. Be sure to periodically review the latest printing of the DOT/RSPA Emergency Response Guidebook and note in the Table of Placards and Initial Response Guides that there are several other hazard classes that also present ignitable hazards. DOT Hazard Classes 2, 3, 4 and 5 all contain materials having properties as ignitable hazards.

TOXIC AND HEALTH HAZARDS

Materials having toxic or health hazards are characterized by having a potential to make people sick, injured, or even cause death. Health hazard severity and death potential are viewed from acute and chronic perspectives. Figure 3.9 denotes a comparison between acute and chronic health issues.

HUMAN HEALTH CONSIDERATIONS

Unfortunately, occupational safety and health standards indicate protecting the "average" worker. In this regard, would an average person please step forward? Many experienced environmental health and safety management professionals agree that every person has an individual threshold limit that must be triggered before negative health symptoms are manifested. Some people may develop symptoms from a microgram of hazmat and others may require a dump truck full. Acute or chronic health conditions arise from being subjected to the "Four Modes of Exposure" as shown in Figure 3.10.

Please know that the author encourages you to keep in mind resources for finding information as you work in the environmental, health, and safety careers. However, there are some topics that should be memorized. Know properties listed in the four modes listed in Figure 3.10. They are integral to working safely, personal protection, and staying alive. We call these kinds of topics critical competencies.

Figure 3.8 DOT Hazard Class 6 - Emergency Response Guide 153.

Critical Competencies

Critical competencies must be learned and always kept in mind to ensure that proper standard operating procedures are implemented and followed by hazmat workers. Methods of personal protection, knowing when to stop, and when to evacuate also reflect upon understanding critical competencies. What about common sense? Knowledge is regarded as listening and reading between the lines; experience is wishing that you did. Common sense is finding a way to understand the intent of what was spoken, what was written, and what was seen. Remember, there are no old bold pilots, nor any old bold hazmat workers. Hazmat safety is usually not depicted correctly in the movies.

HEALTH CONDITIONS RELATED TO DANGEROUS SUBSTANCE EXPOSURE

When proper standard operating procedures are not followed and proper protective measures are not taken, exposure to dangerous substances will cause acute or chronic health conditions and sometimes death. Manufacturers of dangerous substances provide material safety data sheets (MSDS) to warn anyone that may work with or around a dangerous substance. The federal Occupational Safety and Health Administration (OSHA) and every state safety and health department requires each employer to ensure that every affected employee is properly trained to work with

ACUTE HEALTH HAZARDS ──────────▶Rapid Onset of Symptoms

CHRONIC HEALTH HAZARDS────────▶Symptoms After Extended Time

Figure 3.9 Comparison between acute and chronic health issues.

Inhalation -- Exposing the respiratory system: the fastest mode for entering the body.
Ingestion -- Exposure through eating: Entry by utilizing the digestive system.
Absorption -- Transfer through skin: Entry through infinite amounts of skin surface area.
Injection -- Exposure through wounds, open skin, and punctures: Entry into the circulatory system, another quick mode of entry into the body.

Figure 3.10 Four modes of hazmat exposure.

and around hazardous materials. The OSHA regulation 29 CFR 1910.1200, Hazardous Materials Communication (HAZCOMM) requires an ongoing program to ensure safe handling and prevention of exposure to dangerous substances. The following examples are health conditions related improper chemical handling.

Irritation

Irritation can involve skin, eyes, and respiratory conditions related to varieties of chemical exposure. Examples may include organic solvents that remove oils from tissues, cracking of skin, and aggravation of mucous membranes. Also, corrosive liquids, vapors, gases, and most other hazardous substances can produce irritation.

Synergism

Exposure to a specific chemical may produce unique symptoms. However when coupled with other substances, symptoms may change. In this regard, singular exposures to a specific substance may not cause serious health effects. When combined with exposures to other substances, a specific "moderately hazardous" substance may take your life (e.g. alcoholic beverages plus barbiturates).

Sensitization

Most people are exposed to materials that trigger a "memory effect" upon their immune system. Unfortunately, we may not know when this occurs. On future exposure occasions to these memory-making materials we may see some negative health effects. Examples may include hives, dermatitis, anaphylaxis, and even death.

Asphyxiation

In most cases, substances that cause asphyxiation interfere with the oxidation processes within living tissues. Products that cause asphyxiation are characterized as following:

- Simple asphyxiants may include gaseous substances such as carbon dioxide, nitrogen, hydrogen, noble gases (xenon, krypton, argon, neon, and helium), and saturated hydrocarbons (such as natural gas, methane, ethane, propane, and butane). Simple asphyxiants can dilute or displace the oxygen in our breathing air. Acceptable oxygen levels in normal environments range from 19.5 % to 21% and 20.8% oxygen content in breathing air is considered optimum.
- Blood asphyxiants are substances that combine with the red blood cells and alter cellular properties required to form oxyhemoglobin, which is the mode of carrying oxygen to living cells throughout the body. Blood asphyxiants include carbon monoxide, aniline, and nitrobenzene.

- Tissue asphyxiants are substances that are carried by the red blood cells to the tissues and then attach themselves to oxygen receptor sites within tissue cells, blocking needed oxygen from the tissues. Cyanide is a common tissue asphyxiant.

Respiratory Paralysis

Respiratory paralysis can be caused by dysfunction of the respiratory nervous system when certain dangerous materials enter the body. These products have an effect upon the olfactory nerve and the presence of dangerous odors is not transmitted to the brain. Basically, the sense of smell is eliminated when the respiratory system is subjected to a respiratory paralyzing substance. Respiratory paralyzing substances include hydrogen sulfide, carbon disulfide, acetylene, ethylene, diethyl ether, acetone, and ethanol. Many of these products have an anesthetic or narcotic effect upon the body.

Systemic Poisoning

Systemic poisoning is caused from exposure to substances that interfere with the function of major body processes. Modes of systemic poisoning are characterized by specific organs and systems related to proper functioning of an organ as follows:

- Dysfunction of liver and kidneys can be caused by hazardous exposures to products such as arsenic, lead, mercury, and cadmium. Also halogenated hydrocarbons are considered to produce negative affects upon the liver and kidneys, as well as upon other organs.
- Bone marrow illness (including anemia and leukemia) can be caused by hazardous exposures to products such as benzene and suspect substances as toluene, xylene, and naphthalene.
- Muscle disorders can be caused by products such as strychnine. In cases involving severe exposure, victims experience extreme muscle contractions - some causing breakage of most bones in the body.
- Nerve dysfunction can be caused by products that interfere with vital nerve impulses. In cases involving significant exposure, death will result after a series of painful events. Substances that cause illness to the nervous system may include materials such as organic phosphates, carbon disulfide, and methanol.

Mutagenesis

Mutagenesis is a process that can be initiated by mutagens which may include dangerous materials, microbial agents, or electromagnetic energies that change genetic information (DNA) within human, animal, or plant cells. In some cases mutagenesis can be desirable if the end product positively supports overall life properties of the host. Unfortunately, a wide variety of negative health conditions can result in mutated cells such as cancer or unnatural genetic-related health conditions which may surface later in life. Always check in material safety data sheets (MSDS) for mutagenic potentials of a product prior to working with the product. Suspect mutagen sources are too many to list, especially chemical and environmental mutagens. This area of study is so vast that associations and societies have been established to consider negative potentials of environmental mutagens, for example, Environmental Mutagen Society (EMS).

Teratogenicity

The ability to cause birth defects by exposing an expecting mother and her unborn child to teratogenic materials is called teratogenicity. Teratogens are chemical or physical agents that can lead to malformations of the newly forming cells of unborn children. Teratogens may be found in drugs, industrial chemicals, as well as consumer products. The world was shocked during the 1950's and 1960's when thousands of children were born with horrifying birth defects caused by a drug prescribed by obstetricians that contained the product thalidomide. As mentioned about mutagens, there are hundreds of substances that are considered to be teratogenic. Consult MSDS and other technical information to determine teratogenic properties of a substance before using it.

The following are few examples of many suspected teratogens:

- Heavy metals inhibit enzyme activity and causes defects in the central nervous system (CNS).
- Tobacco smoke can cause the same health conditions that affect the mother, such as heart disease, cancer, and organ malfunctions.
- Alcohol can cause learning disabilities and psychomotor dysfunction.
- Actinomycin inhibits formation of genetic materials (RNA/DNA).
- Alkylating agents interrupts DNA synthesis and causes mutations.

Carcinogenesis

Carcinogenesis is a cellular process that may involve chemicals, microbial agents, or electromagnetic energies to form a tumor or specifically cancer. Human cancer is a condition that involves unchecked, uncontrolled, or unregulated proliferation of human cells - a condition often referred to uncontrolled mitosis or cell splitting. It is estimated by some researchers that approximately one American out of four will experience some form of cancer in their lives. Unfortunately, we do not fully understand why some chemicals cause cancers in some individuals and why not in others. Cancers are characterized by the types of tissue or cells that may be affected such as:

- Leukemia involves the bone marrow and formation of white blood cells.
- Lymphomas involve tissue of the lymphatic system. A common disease that is related to lymphatic cancer is Hodgkin's disease.

- Sarcomas involve connective tissues, bones, and cartilage
- Carcinomas involve protective surfaces outside of the body, coverings of internal organs, mucosal tissues, respiratory system, digestive tract, and others. Carcinoma is considered the most common form of cancer.

Carcinogens are characterized in two categories. Genotoxic carcinogens are substances that interact with cellular genetic codes. These carcinogens are considered as initiators and are primary, direct-acting carcinogens. They may be associated with:

- Synthetic plastics
- Coal tar (benzopyrene)
- Nitrosamines
- Metals (especially heavy metals and cadmium, nickel and chromium)
- Cigarette smoke (N-dimethylnitrosamine and others)
- Mustard gas

Epigenetic carcinogens are considered promoters that enhance the growth of pre-cancerous cells within target tissues. These carcinogens may be associated with:

- Asbestos
- Beryllium
- Estrogens and androgens
- Ethyl alcohol
- Azathioprine
- Many organic solvents
- Tetrachloroethylene, also called perchloroethylene, used in chemical processing of printing plates and for dry cleaning clothing.

SYSTEMIC TOXICITY

A substance is considered toxic by its ability to cause harmful health effects upon a living system. Key factors relating to toxicity are dosage or amount of toxicant and the time of exposure to body systems. The degree of harm caused by a substance is viewed from short-term (acute) and long-term (chronic) perspectives. When the substance is distributed throughout the body and when eleven key organ systems are affected, relative health issues are considered systemic.

Organ Systems

The human body contains eleven organ systems that are easily affected by exposures to dangerous substances:

1. Skin
2. Eyes
3. Blood (plasma, platelets, red and white blood cells)
4. Respiratory system
5. Circulatory system (cardiovascular and lymphatic)
6. Endocrine system (pituitary gland, thyroid gland, adrenal gland, pancreas, testis, ovaries)
7. Liver
8. Kidneys
9. Reproductive system (male and female genitourinary systems)
10. Nervous System (central and peripheral systems)
11. Immune system

Toxic Substances

The following are examples of toxic substances:

- Pesticides including insecticides, herbicides, fungicides, and rodenticides
- Fumigants
- Metals
- Solvents
- Animal Toxins
- Plant Toxins
- Microbial Toxins

Typical Pesticides

These are dangerous materials that are added to the environment for the expressed purpose of killing life forms that are vexing to humans. Acute and chronic exposures to pesticides frequently occur in humans from improper handling, unauthorized use, and not using proper personal protective equipment. In most cases, toxic exposures can occur from any of the four modes of exposure: inhalation, ingestion, absorption, or injection which includes puncture, entry into open wounds and cracks in the skin. Because of inherent hazards related to handling pesticides, most states require special licensing of properly trained, authorized persons that may apply pesticides within industry or within the public sector. The United State Environmental Protection Agency now regulates insecticides, fungicides, and rodenticides under the FIFRA environmental act.

Insecticides

Typical substances used in manufacturing insecticides include:

- Organic-phosphorus insecticides, which include substances used in chemical warfare such as: Dimethyl phosphonoamidocyanidate (also called tabun); Isopropyl methyl-phosphonofluoridate (also called sarin); Tetraethyl pyrophosphate (TEPP); and Parathion.

- Other toxic organophosphate insecticides include mevinphos, disulfoton, chlorfenvinphos, diazinon, dimethoate, trichlorfon, and chlorothion. Organophosphates typically affect the nervous systems and can cause respiratory failure.

- Carbamate insecticides act very similar to organophosphates; however they usually demonstrate low dermal toxicity. One of the least toxic carbamate insecticides, Carbaryl, is considered to have teratogenic properties and will likely cause birth defects. Other carbamate insecticides include Baygon, Propoxur, Mobam, Temik, Aldicarb and Zechtran. Typical toxic symptoms from overexposure to carbamates include eye irritation and tearing, salivation, pupil contraction, convulsion, and death.

- Organochlorine insecticides include DDT and cyclodeine chemicals such as chlorodane, aldrin, dieldrin, epatachlor, and toxaphene. Also included in this group are the hexachlorocyclohexanes such as lindane. This family of insecticides can remain in the environment and will bio-accumulate. Because of their persistence, they can produce increased toxicity in affected living systems. They are considered neurotoxins.

- Botanical insecticides include substances such as nicotine and a chemical group called the rotenoids. That includes pyrethrum which is found in several household insecticides. They have the ability to rapidly act upon insect nervous systems and are highly toxic to fish and other aquatic life.

Herbicides

These chemicals destroy plants; however they are also considered dangerous substances to humans. The following are chemical families are considered herbicides:

- Chlorophenoxy compounds, which include dichloro (2, 4-D) and trichlorophenoxyacetic acid (2, 4, 5-T), are commonly used to control woody plants. These substances produce a variety of effects upon humans such as paralysis and heart attack. Skin contact produces chloracne in humans and products that contain these substances are also considered teratogenic. These compounds combined with dioxins are known by Vietnam veterans as "Agent Orange." This dangerous substance is known to cause liver damage, affect human metabolism, cause birth defects (teratogen), genitourinary and reproduction disorders, immune suppression, tissue breakdown (necrosis), and tumor formation.

- Dinitrophenol is known to cause increases in human body temperature. Compounds of dinitro-orthocresol are suspected to cause chronic increases in metabolic rates and changes in the eyes which include the development of cataracts.

- Biparietal herbicides include paraquat which can cause lung damage in humans. A similar compound, diquat, targets the digestive system and damages internals organs.
- Carbamate herbicides include the products propham and barban, which at this writing appear to produce minor toxic symptoms in humans.

Fungicides
These chemicals are used to kill fungi such as molds, yeasts, and other mycotic organisms. Some fungicides contain mercury compounds that are very toxic in humans. Other fungicides such as Captan have shown potential to be teratogenic, mutagenic, and carcinogenic in humans. Aromatic fungicides such as Pentachlorophenol (PCP) and Pentachloronitrobenzene (PCNB) are very commonly used today; however, if improperly used, they can cause systemic poisoning in humans. Dithiocarbamate fungicides have low human toxicities; however lab tests are now indicating potentials for teratogenic-related and carcinogenic-related diseases in humans. Nitrogen heterocyclic fungicides such as thiabendazole and benomyl are suspected in causing dermatitis in workers.

Rodenticides
The function of rodenticides is to control rodent infestations. The prime exposure hazard for humans is acute poisoning due to improper use and handling. Some common rodenticides include:

- Warfarin: This product is commonly used for poisoning rats. It is an anticoagulant and victims of exposure will suffer internal hemorrhaging.
- Red Squill: This is plant plant glycosides which cause irregular heartbeat and eventual death.
- Norbormide: A chemical that causes smooth muscles to contract and stop blood flow.
- Sodium Fluoroacetate: This chemical causes blockage to the Krebs cycle, which inhibits cellular metabolism and death.
- Strychnine: This chemical is a nerve toxin (neurotoxin) that causes convulsions.
- Zinc Phosphide: A chemical that causes severe, debilitating gastrointestinal spasms.

Fumigants
These are pesticides that are released as a gas or vapor. Widespread penetration of toxic agents into inaccessible locations is the prime value of fumigant products. Inhalation is the main human exposure hazard for fumigants. Some common pesticide fumigants include: acrylonitrile, carbon disulfide, carbon tetrachloride, chloropicrin, ethylene dibromide, ethylene oxide, hydrogen cyanide, methyl bromide, and phosphine.

Metals
Metals entering the body in quantities greater than they may be used for normal processes are considered toxic. Even though increases in concentrations of most metals cause negative health effects, great consideration should be given to the heavy metals such as lead, mercury, silver, chromium, cadmium, selenium, arsenic and others. Most increased concentrations of heavy metals affect the nervous system. Other negative health conditions are also manifested in abnormal function of the internal organs in the body.

Solvents
Organic solvents and solvent mixtures readily enter the human body by the four modes of exposure. Due to properties of volatilization, vapor density, and often lack of concern by solvent handlers, acute and chronic solvent health affects become systemic. Typical solvents that are considered health hazardous include:

- Aromatic hydrocarbons, which include benzene, alkylbenzenes (toluene, ethyl benzene, cumene, and xylene), cause internal organic damage, skin disorders, damage to the central nervous system, blood disorders, cancer, and reproductive system abnormalities.
- Chlorinated aliphatic hydrocarbons, which include chloroform and carbon tetrachloride, cause internal organ damage, cardiopulmonary disorders, effects upon the nervous system, cancer, and other health problems.
- Halogenated alkanes and alkenes include vinyl chloride, fluoride and bromide, methylene chloride, trichloroethylene, trichloroethane and similar solvents. Most health effects involve damage to internal organs, cancer, damage to the central nervous system, and other serious health concerns.
- Aliphatic alcohols include ethanol, methanol, glycols, and glycol ethers. Most of these products cause nervous system depression. Some target the retina and damage vision, and most damage internal organs, affect the genitourinary system, and some are teratogens.
- Aliphatic hydrocarbons include natural gas, methane, ethane, propane, butane, hexane, hexanone (methyl ethyl ketone, MNBK), gasoline, and kerosene. Along with presenting extremes of fire hazards, some cause asphyxiation, central nervous system depression, damage to internal organs, and sensory loss in the extremities.
- Carbon disulfide, which is commonly used in plastics manufacturing, causes damage to vision and hearing, and contributes to heart disease.

Animal and Insect Toxins
Poisonous and venomous animals and insects are considered hazardous and toxins that they produce are considered dangerous to humans. Assessing hazardous conditions must always include being aware of the presence of hazardous species of animals and insects such as:

- Reptiles, such as lizards and snakes
- Amphibians, including frogs, salamanders, and newts
- Marine animals such as protista, radiate, acoelomata, and pseudocoelomata
- Gastropods such as snails, slugs, cone shells, and limpets
- Bivalves such as certain types of shellfish
- Fish such as stingrays, stonefish, jellyfish, and several others
- Arthropods such as spiders, scorpions, centipedes, ants, bees, ticks, fleas, and many others

Human health effects associated from poisonous bites will vary; however, it must be known that serious health reactions and even death potentials must assessed from these sources in most environments that humans may exist.

Plant Toxins

Toxins associated with plants may cause human health effects that range from gastric disturbance, convulsion, neurological disorders, skin disorders, systemic poisoning, and death. In some instances plant toxins have been developed into substances that are used in drugs for medical treatment. Unfortunately, some have been converted into weapons of mass destruction.

REACTIVE HAZARDS

Some substances undergo physical reactions that will release uncontrolled energy. In some environments reactive substances become unstable or may react violently in the presence of other materials or physical conditions. Some reactive substances can release extreme heat, fire, or explosion, and also release toxic emissions. They may react as the environmental conditions change such as temperature extremes, relative humidity, shock, or friction. An explosive condition that is often overlooked when dealing with heat initiated reactions is called BLEVE or boiling liquid expanding vapor explosion. When exposed to heat, contained materials will expand and the container may rupture or even explode. When the contents are ignitable, the BLEVE may yield disastrous results. Most firefighters do not hastily enter locations where containers of solvents, gases, and other expandable materials are stored. Contained water is even subject to BLEVE. Unfortunately many people have witnessed an exploding hot water heater.

Is water considered a reactive hazard? Not by itself, but if combined with water reactive substances, it also takes part in some violent reactions. Let's review the kinds of substances that can be categorized as reactive.

Unstable Substances

When we consider unstable materials we must include organic peroxides and monomers. Unstable materials demonstrate properties such as spontaneous decomposition, polymerization, and other means of auto-reaction.

Monomers

Monomers are simple organic molecules that can self-react when exposed to heat or other physical conditions and grow into giant molecules called polymers. Most monomers are flammable gases that are easily liquefied. Unfortunately, in conditions involving high heat, monomers produce a severe reaction called runaway polymerization, also called uncontrolled polymerization, which often leads to BLEVE and severe explosion. Monomers are characterized as potentially unstable and are usually stored away from normal flammable substances. Chemically, most monomers are unsaturated hydrocarbons that contain one double bond. When exposed to heat and pressure the double bond is released allowing openings at the ends of the molecule, sometimes called dangling bonds that react to join the monomer molecules together as one large molecule. In uncontrolled conditions this reaction can yield high amounts of released energy and explosion. Under controlled conditions, this same reaction is used to form thermo-plastics. All monomers are not exactly alike and there are many.

Hazardous properties of monomers include being:

- Unstable
- Often toxic
- Often corrosive
- Sometimes under pressure
- Extremely reactive
- Uncontrollably ignitable and flammable
- Easy to produce BLEVE

Organic Peroxides

Organic peroxides are organic substances that contain enough oxygen that is easily released to promote oxidation, fire, and or explosion. A considerable characteristic of organic peroxides is the absorption of energy that serves as an internal ignition source. At ambient temperatures, most organic peroxides remain somewhat stable. However as temperatures increase, the molecules disintegrate and energy is released. The temperature that initiates absorption of energy into an organic peroxide molecule is called SADT or self-accelerating decomposition temperature. Once molecular decomposition begins, it cannot be reversed.

There are many organic peroxides and they are most apparent by chemical names that contain "ate," "ide," "oxy," "per." The following examples are typical organic peroxides:

- Acetyl peroxide
- Dibenzoyl peroxide
- t-butyl peracetate
- Dicumyl peroxide
- Peracetic acid

- Peroxyacetic acid
- Methyl ethyl ketone peroxide
- Methyl isobutyl ketone peroxide

Hazardous properties of organic peroxides include being:

- Unstable and are oxidizers
- Uncontrollably ignitable and flammable
- Extremely reactive
- Often explosive
- Often corrosive
- Often toxic

Water Reactives
Many substances will react violently when coming in contact with water. Some materials will emit flammable gas that is often ignited by heat generated by exothermic reaction or in the presence of other ignition sources. There are many water reactive substances and we will discuss a variety of them. One grouping of water reactive materials is the alkali metals.

Alkali Metals
Alkali metals include lithium (Li), sodium (Na), potassium (K), rubidium (Rb), and cesium (Cs). These materials will all react violently with water by rapidly separating oxygen from hydrogen and, due to exothermic reaction, hydrogen is ignited during the same reaction. The alkali metal forms hydroxide (OH) and a corrosive alkali substance is formed.

Alkaline Earth Metals
Alkaline earth metals are somewhat less reactive than the alkali metals. They include beryllium (Be), magnesium (Mg), calcium (Ca), strontium (Sr), barium (Ba), and radium (Ra). These metals are often used in industrial processes and some products are made from them. These include lightweight magnesium products and beryllium copper alloys. Barium and strontium are used in electrical parts. Radium is radioactive and usually used in small quantity. Small pieces, grindings, or dusts violently react, spontaneously ignite, and also become pyrophoric. Once burning, fire involving alkali metals or alkaline earth metals is extremely difficult to extinguish. Type D fire extinguishing agent is used to encapsulate the burning particles. Other means of extinguishing will scatter the burning parts and, in some cases, burning will violently increase.

Hydrides
Hydrides are inorganic materials combined with hydrogen that violently react with water. When exposed to water, they release hydrogen and form hydroxides.

Hazardous properties of hydrides include irritation, toxicity, and extremes of flammability. When exposed to water they form corrosive caustic compounds and liberate explosive quantities of hydrogen. Typical hydrides include lithium hydride (LiH), sodium hydride (NaH), and boron hydrides (BH_3 or Borane). Borane is unstable at ambient temperatures and will break down to become other products that are extremely flammable or explosive gases that are extremely toxic.

Carbides

Carbides are carbon-containing compounds and some of them react with water to produce flammable gases and caustic compounds. One commonly known is calcium carbide (CaC_2), which has an odor similar to garlic and, when exposed to water, forms a strong caustic compound Calcium hydroxide (CaOH) and liberates an extremely unstable and flammable gas, acetylene. Other water reactive carbides include aluminum carbide (Al_4C_3) which liberates methane, beryllium carbide (Be_2C) which also liberates methane, and magnesium carbide (Mg_2C_3) which liberates propyne (methyl acetylene) gas which is a key component of MAPP gas. MAPP gas is used for high temperature welding and includes methyl acetylene, propadiene, propane, and butane.

Nitrides

Nitrides are water reactive compounds that contain nitrogen as a nitride ion N_{-3}. Examples of water reactive nitrides are magnesium nitride (Mg_3N_2) when combined with water forms ammonia (NH_3) and caustic magnesium hydroxide (Mg [OH] $_2$). Lithium nitride (Li_3N) is water reactive and pyrophoric. When combined with water it forms ammonia and caustic lithium hydroxide. Ammonia when combined with water forms corrosive ammonium hydroxide (NH_4OH).

Phosphides

Phosphides are water reactive solid compounds containing the phosphorus ion (P_{-3}). When phosphides react with water, they form flammable, toxic gas phosphine (PH_3) and a caustic hydroxide. Aluminum phosphide (AlP) when combined with water forms toxic gas that is used as a fumigant insecticide. Calcium phosphide (Ca_3P_2) is dangerous when wet and is used as a pyrotechnic and a rodenticide.

Inorganic Chlorides

Inorganic chlorides react with water, and many only dissolve in water. However some will violently react when exposed to water to generate toxic acid chloride fumes and corrosive acid chloride liquids.

Peroxides

Peroxides including inorganic peroxides, sodium peroxide, lithium peroxide, and potassium peroxide, produce hydrogen peroxide when exposed to cold water. Depending upon increased water temperature, hydrogen and oxygen may be liberated.

Other Significant Water Reactives
Other significant water reactives include:

- Calcium oxide (CaO) - produces an exothermic reaction and formation of calcium hydroxide which is a caustic corrosive alkaline
- Sulfuric acid (H_2SO_4) - produces a violently rapid exothermic release of energy
- Acetic anhydride ($(CH_3CO_2)_2O$ - liberates concentrated acetic acid fumes when mixed with water
- Acetyl Chloride (CH_3COCl) - liberates hydrogen chloride gas (HCl) and acetic acid when mixed with water

Air Reactives
Pyrophoric substances react spontaneously in air, often burst into flame, emit toxic fumes, and in some cases, may detonate. Two common categories of pyrophoric substances are phosphorus and some organo-metallic compounds. Phosphorus is white or yellow and is usually kept underwater. It burns violently upon drying. The military often uses white phosphorus in tracer bullets and grenades. When exposed to air, each piece will burn through metal and exposed human flesh. Red phosphorus is not pyrophoric. As a process mechanism, white phosphorus is chemically treated to become red phosphorus to reduce safe handling liabilities and then is reversed back to white or yellow phosphorus when desired.

Many organo-metallic compounds are pyrophoric. These materials have organic radicals that are affixed to a metal. They are named using typical organic chemistry and metallic nomenclatures. For example, dimethyl arsine ($[CH_3]AsH$) is a colorless toxic liquid that is pyrophoric. Triethyl aluminum, trimethyl aluminum, trimethyl aluminum ethereate, chlorodiethyl aluminum, and tri(iso)-butyl aluminum are clear liquids that ignite in air and will explode if contacted with cold water. Also, dimethyl cadmium and diethyl zinc are pyrophoric.

When alkaline earth metals and alkali metals are reduced to fine powders, they also react with air. It is believed that moisture in air initiates reactions with these metals. Caution is stressed in choosing the proper extinguishing agent when these materials are involved in fire. Most standard extinguishing agents will cause pyrophoric agents to violently react.

BIOLOGICAL HAZARDS

Biohazards are pathogenic microorganisms that are capable of causing disease in humans and other animals. Biohazards may include bacteria, fungi, protozoa, viruses, and other microbial life. Later in this text we will discuss using the aseptic procedure for controlling potentials for exposure. We will also characterize variables relating to microbial size, chemistry, life cycles, and the microbial production of dangerous substances that cause disease. Unfortunately, we will also need to discuss the involvement of biohazards in weapons of mass destruction, a threat to our world and to the world of future generations. The objective of this section is to characterize hazardous properties of biohazards and exposure potentials from normal environmental sources.

Considerable Sources of Biohazards

From the minute of our birth we are exposed to hundreds of microbial agents that can hinder our lives. Every day we battle organisms that exist in the environment, within and outside the bodies of other people, and especially those that inhabit human and animal body fluids, feces, excretions, and more. Bloodborne pathogens are organisms that are found directly in blood, blood products, and other human body fluids. These would include the AIDS virus (Human Immunodeficiency Virus [HIV]), hepatitis, and several others. Another concern involves pathogens that exist in the environment, especially in damp or wet places. Much media coverage has focused upon indoor air quality and human living spaces, especially waterborne microbes such as mold, legionella and others.

How do humans become exposed to disease-causing microorganisms? Remember the modes of exposure that we covered during discussions of health hazards? We are exposed to pathogenic microbes in the same manner as we are exposed to other hazardous substances, by absorption through the skin, by inhalation, ingestion, and through skin punctures, cuts, and other invasion portals that lead to our circulatory system. Due to exposure fears related to AIDS and hepatitis, the U.S. Occupational Safety and Health Administration (OSHA) promulgated the Bloodborne Pathogens Standard for workers that work in potentially infectious environments.

Blood and Body Fluids
Whenever situations arise that involve human body fluids, we should be concerned with exposures to the HIV and varying degrees of hepatitis. To date, AIDS infection or specific varieties of hepatitis cannot be completely cured. In this regard, all exposures to human body fluids must be considered potentials for serious disease. Handling of human body fluids must follow aseptic procedures and universal precautions. Aseptic procedure involves using personal protective equipment and following methods that prevent exposure to disease-causing viruses. Please remember that viruses are considered obligate parasites and they need living cells to remain alive themselves. Anytime that blood or body fluids are "fresh and wet" they most likely contain living cells that host living viruses. Disinfection of contaminated surfaces and materials must be complete. In most cases, disinfection involves washing with lots of soap and water and final rinsing with an acceptable disinfectant that penetrates remaining residues. Disinfecting areas of microbial contamination involves removal of chemical toxins generated by microbial activity and killing remaining "live" organisms. What are acceptable disinfectants? Some professionals use proprietary preparations that are guaranteed to kill pathogens; others may use chlorine bleach solutions that contain at least 100 parts per million active chlorine. Still oth-

ers may use quaternary ammonium solutions in concentrations of at least 200 parts per million. Please be aware that most "off the shelf" spray disinfectants have very limited success in disinfecting contamination from dangerous pathogens.

Animals
Animal sources can be wild or domesticated. Where does anthrax come from? Another name for anthrax is "wool sorters disease." The intestinal tracts of most animals contain live bacteria and viruses. How do infectious escherichia coli get into hamburger or steak? E.coli is a normal inhabitant of beef intestines. Mad cow disease? We're not even going there right now. You may want to have second thoughts about eating "rare," almost uncooked, meat. What do you know about the turkey trots? Not the dance, the salmonella toxin induced stomach distress that often occurs after eating stuffing that was warmed in the intestinal tract of a Thanksgiving bird. By the way, eggs are used in the lab to grow viruses. We could go on and on, but we need to eat something, so this may be good place to hold on this subject for now.

Dusts and Spores
One Sunday afternoon between football games, you may have noticed little specks of materials floating in the sunlight that is beaming through your window. What are the specks? Dusts and spores. What does dust contain? Let your mind wander but please include fine particles of insect and rodent feces and lots of other stuff. Spores are varieties of bacteria and fungi that have "gone to sleep" in order to adapt to environmental changes. Upon drying, many microbes slow down all biological functions and form a barrier that is resistant to extremes of temperature and drying. When conditions occur that promotes routine metabolic activity such as moisture, temperature, proper oxygen or carbon dioxide content, and food sources, the microbe awakens to regular activity. Bacteria that cause tetanus, botulism, anthrax and many other diseases function in this manner.

Fungi, such as many hundreds of molds, form spores in order to spread throughout the environment. By the way, black mold is not the only species of mold that produces spores, that spreads throughout the environment, and produces active cells that cause health problems caused from fungal mycotoxins. It is questionable, however, that mold causes death in healthy humans. In following sections of this text, we will discuss other concerns related to biohazards such as safe working conditions, proper handling procedures, and methods of decontamination.

RADIATION HAZARDS

Radiation hazards are classified as ionizing and non-ionizing. Ionizing radiations will penetrate the body, whereas non-ionizing radiations will only have an effect upon body surfaces and the eyes.

Non-Ionizing Radiation Hazards

Ultraviolet light and heat from infrared radiations are most commonly associated with non-ionizing radiations. Light energies generated from the sun as well as energies generated from electromagnetic radiation sources such as laser light are examples of ultraviolet radiations.

Non-ionizing radiation involves radiant heat, radio waves, ultraviolet light, and light. Sources may include laser light, ultraviolet light from welding, the sun, sunlamps, ultraviolet light sources used in controlling microbiological life in air and water systems and others. N.I. radiation is responsible for thousands of skin cancer victims as well as eye and vision impairments.

Ionizing Radiation Hazards

Penetrating radiation is energy that is emitted from nuclear interactions as sub-atomic particles and high-energy photons are released as radioactive decay. Some naturally existing substances continually undergo radioactive decay such as unstable isotopes of radium, radon, uranium, and thorium. Other substances undergo man-induced nuclear reactions. Three of the most common types of ionizing radiation are:

- Alpha radiation: The largest particle emitted from an atom. Because of the large particle size, alpha radiation energy is quickly depleted due to many collisions with other atoms. Alpha particles will only travel a few centimeters in air. Most substances will provide a protective barrier to prevent hazardous exposures to alpha particles. *Caution: Hazardous exposures to alpha radiation* include inhalation and ingestion. If alpha radiation enters the body, the mucosal tissues and internal organs can be damaged.
- Beta radiation: A smaller particle that collides with fewer atoms than alpha particles. Because of this property, they can penetrate more materials include human tissues. Thin pieces of metal, plastic, glass and thick pieces of wood will contain beta radiation. Beta radiation can damage the skin, eyes, and internal organs if ingested or inhaled.
- Gamma radiation: A radiation that is considered "pure energy," such that it does not have mass or charge of attraction. Gamma radiation will penetrate most materials as energy waves, including the human body. Human cells will undergo physical damage and mutation due to exposure to gamma radiation. Dense materials such as lead and several feet of cement are not penetrated by gamma radiation.

Exposure to Radiation

Human cells are damaged by penetrating radiation. It is noted that rapid replicating cells show the greatest response, or radio-sensitivity, to exposures of radioactivity. Human cells that exhibit rapid replication, cell splitting, or mitosis, are:

- Transitional white blood cells
- Cells of the reproductive system including sperm and ovum
- Digestive system cells
- Blood producing organs and associated cells
- Rapid-forming cells of unborn children

Depending upon individual health, age, and overall physical condition, human cells that resist damage from radiation are connective tissues such as bones and muscles, and nerve cells.

Acute Exposures to Radiation
Large doses of radiation within short periods of time, usually in excess of 25 rem dosages are considered acute exposures. Symptoms of acute radiation exposure may include:

- Decreased numbers of white blood cells
- Internal bleeding and shock
- Nausea and vomiting
- Diarrhea
- Fatigue
- Poor appetite
- Microbial infections
- Discoloring and redness of skin
- Hair loss
- Sexual sterility
- Convulsions
- Coma and death

Chronic Exposures to Radiation
Exposures to low levels of radiation over long periods of time are responsible for health effects such as cancer, cataracts, and abnormalities of the genitor-urinary and reproductive system. Unfortunately, negative genetic effects may be passed on for several generations.

Methods of handling and decontaminating radioactive materials will be covered later in this text. However, please remember three principles of safety when dealing with potential exposures to radioactivity, these considerations will be discussed again:

1. Time: Less time of radioactive exposure yields less potential for health affects.
2. Distance: Farther the distance from radioactive sources yields less exposure.
3. Shielding: Place protective materials between you and radioactive sources.

Chapter 4

HAZARDS CHARACTERIZATION AND SITE EVALUATION

The Hazards Categorization (HAZCAT) approach is covered in this chapter. Every worksite contains hazards; unfortunately some hazards are somewhat obscure. It's up to the environmental, health, and safety professional (EHS) to find all site hazards, as well as any condition that is out of regulatory compliance. This type of work requires dedicated responsibility, involves professional liability, and a screw up could cause a lot of damage or, worse yet, loss of life. One of the first disciplines learned by effective environmental, health, and safety professionals is hazards categorization or HAZCAT. Environmental, health, and safety work often takes us into the realm of conducting environmental and safety site audits and assessments.

The HAZCAT approach requires EHS professionals to use common sense, which is a prime requirement for anyone that claims to be a certified environmental auditor and assessor. Hazards categorization cautions environmental, health, and safety professionals to conduct site evaluations with an open mind and to:

1. BE SAFE
2. BE SURE
3. BE SUCCINCT

Safe work practices must be imbedded into standard operating procedures for anyone that recommends, oversees, or conducts operations in potentially hazardous environments. Ensured decisions must be made using scientific proof and known information. To be succinct, anyone conducting, or directing potentially hazardous operations better be 100% accurate, or have a very good excuse if not. Luck is not a feasible, legal, or moral entity.

THE PROFESSIONAL

Note that the term "certified" is mentioned above. According to the dictionary, the term certify means to "attest to truth and validity." People that are certified to perform hazardous operations must be tested; their knowledge and abilities must be proven worthy, and how they were proven worthy must be valid. Professionals that are not exactly certified as an environmental, health, and safety (including hazardous materials) auditor or assessor should not be performing EHS audits or evaluation functions. For more information on this subject, contact organizations such as the National Registry of Environmental Professionals (NREP), American Society of Safety Engineers (ASSE), the National Environmental, Health, and Training Association (NEHSTA), the Regional OSHA Training Institute Satellite in your area.

SITE CATEGORIZATION

Conditions will vary, as well as requirements to expediently assess hazard potentials as situations may vary. HAZCAT procedures performed during emergency conditions are important, however, standard operating procedures performed while conducting an environmental audit are important as well. When we discuss emergency actions and response activities in this text, please reflect back upon hazards assessment and sampling procedures mentioned in this chapter.

Someone Must Take Charge

Site evaluators, team leaders, or incident commanders are all responsible to ensure that hazard characterizations and assessments are properly conducted. These same professionals will find themselves answering a wide variety of questions if things go wrong, especially to the media. Emergency responders, biohazard abatement personnel, and mold contractors may have some experience in this regard. This topic will emerge again in later chapters of this text.

Keeping proper documentation for all site hazard or regulatory evaluation activities is of the utmost importance for many reasons. Why document all hazard categorization activities? Proof, follow up, and reminders are all good reasons. Sounds like a good place for checklists? Don't leave the office or truck without them. Please see chain of custody, also called chain of evidence, checklists, and environmental audit checklists later in this chapter.

Routine Site Categorization Activities

In most situations, there are three prime phases of activity initially involved for evaluating potential hazards and determining regulatory issues involved to complete site categorization activities. Certified environmental auditors, assessors, and hazards abatement professionals have learned that this triad of events shall be performed in order to successfully complete an assignment. See Figure 4.1.

Off-Site Survey, Audit, and Evaluation Activities

During emergencies wouldn't it be a good idea to obtain answers from knowledgeable parties before diving into serious conditions and committing personnel, equipment, and resources to unknown situations? Many people have lost their lives because this common sense action did not occur. Likewise, whenever sites require evaluation for hazardous conditions, regulatory issues, or suitability for human habitation, many questions should be answered by knowledgeable parties before committing personnel, equipment, and resources as well.

How can site information be obtained? Review the following procedures:

- Conduct interviews and search records
    1. By conducting interviews with persons that are familiar with the site
    2. By obtaining records such as viewing county court records, real estate documents, records on file with state regulatory agencies and emergency response commissions and others
    3. Obtain aerial maps, meteorological information, and geologic or hydrologic information
    4. Evaluate site layout and terrain
    5. Consider potential modes of dispersing pollutants in air, land, and in waters
    6. Consider locations of populations and potentials for human and ecological risk

Site Categorization Triad
1. Conduct Off-Site Survey and Audit Activities
2. Conduct On-Site Survey and Audit Activities
3. Routine Monitoring of Hazardous Activities

Figure 4.1 Initial site categorization activities.

7. Evaluate accessible means to enter and exit the site by land and air

8. Obtain information on status of activities that have already occurred on site

- Evaluate perimeter and adjacent area interface conditions

    1. Compare visual representations of historical and current aerial photographs and maps

    2. Note from a distance the placards, labels, and other markings on buildings, tanks, pipes, and other containers

    3. Note any occurrence of biological considerations such as dead animals, plant life, and obvious contamination of waters (such as algae in ponds) and obvious chemical contamination (colors, odors, drainage to off site locations)

    4. Monitor air streams at site perimeter and waters draining from the site

    5. Collect and analyze samples taken at perimeter and adjacent locations for chemical and biological hazards in the soil, drinking water, surface waters, ground waters, and ambient air

On-Site Survey, Audit, and Evaluation Activities

Using information obtained from off-site survey, audit, and evaluations, decisions should be made to determine activities allowed to be performed by personnel that enter the site. It must be mentioned at this time that a very important principle must be put in place and kept in place: Never enter a potentially hazardous site alone - never. Under the OSHA HAZWOPER regulation, it is emphasized that the "Buddy System Prevails." Again, never enter a hazardous location alone and always have back up support, or don't go in!

Reconnaissance Personnel

What activities do site reconnaissance personnel perform? First considerations for reconnaissance personnel are protection. The first entry team should always hope for the best and prepare for the worst-case scenarios. If information obtained from off site evaluations are inconclusive or do not prove that hazardous conditions don't exist, first entry personnel may be required to use Levels A or B other maximum protective equipment. Reconnaissance activities may include:

- Determine if any significant hazards exist on site
- Visually locating and documenting physical, chemical, and biological site hazards
- Evaluating potentially emergency conditions
- Noting immediately dangerous to life and health (IDLH) conditions
- Monitoring for airborne hazards in IDLH concentrations
- Monitoring for the presence of ionizing radiation

Second Entry Personnel

Based upon determinations made by reconnaissance personnel, the second entry team may be allowed to enter a site to further characterize an unknown site. The following activities may be performed:

- If no hazards present, a secondary site audit and evaluation would be performed.
- If no hazards found during the secondary audit, other prescribed activities would be allowed to take place.
- If hazards were noted by reconnaissance personnel, the second entry unit may be involved with:
  - Conducting environmental sampling
  - Performing on-site field testing
  - Positioning stationary sampling equipment
  - Installing personnel monitoring devices
  - Conducting ongoing environmental monitoring

Can situations occur that cause changes in site entry activities? Yes, incidents or conditions may warrant a change in game plan. For instance, a new work phase could go into effect if there is:

- An inability to stabilize hazardous conditions found on site
- A change in weather or season
- A change in the scope of the work assignment
- A concern related to administrative issues and others

VALUE OF DOCUMENTATION

As mentioned earlier in this chapter, everything that occurs during site characterization activities must be documented. Documents related to environmental, health, and safety site audits and assessments should be considered controlled. Every operation must have a system for document control. Documents should be kept as hard copy and digital computer records; don't forget to keep backup copies. Yes, common sense must prevail. Good documentation should consist of:

- Data that guarantees accurate communication
- Data from high quality, reliable sources
- Safety and environmental compliance information. If compliance, injury or illness issues arise they must be written, and follow up information must indicate how the concerns were handled or abated. Never leave safety, health, or environmental concerns as open-ended data; always strive to show closure if possible

- Legal and regulatory issues must be documented and, if possible, never left open-ended. Documented legal issues should explain initial concern and current status.

Acceptable Site Characterization Documents

Site characterization documents are controlled documents. Controlled site characterization documents could include the following examples as shown in Figure 4.2.

SITE SAMPLING - PERSONAL SAMPLING

As previously mentioned people are exposed to dangerous substances in four ways. Modes of personal exposure include:

1. Respiratory exposure to dangerous materials, the fastest normal route that substances enter the body. Aerosols enter the lungs and rapidly cross through the alveolar exchange system from outside air into the blood and circulatory system.
2. Absorption through exposed skin.
3. Ingestion that involves eating, or swallowing materials that have been collected in the upper nasal mucosal tissues. Dangerous materials are usually collected as dusts or dangerous properties absorbed into dusts that can be swallowed throughout the day and enter the body through the digestive system. Heavy metals such as lead and mercury often enter the body in this manner.
4. Puncture wounds, sores, and other openings in the skin that allow materials to eventually enter the bloodstream. Microbial infections as well as chemical substances easily enter the body in this manner.

When we enter hazardous environments, we are required to prevent exposures to dangerous substances by using engineering controls or by wearing personal protec-

---

Site Entry Activities
1. Site Reconnaissance
2. Survey, Sampling, and Testing
3. Monitor during Pre-work & Work Activities
4. Work Activities

Figure 4.2 Examples of site survey, audit, and evaluation documents.

tive equipment and by monitoring concentrations of dangerous substances that may exist.

Sampling Techniques

A variety of sampling devices are used to check exposure potentials for personnel that are in a hazardous environment. These devices are grouped into two classifications:

- Passive sampling devices
- Active sampling devices

Most site workers have used handheld, passive, direct reading digital devices that are calibrated to notice the presence of, or lack of, a given substance. Examples may include meters such as:

- Oxygen meters, that are set to read % or parts per million of oxygen. They are often preset to warn personnel if oxygen levels fall below 19.5% or exceed 22%.
- Explosion meters work by comparing normal air environments with airborne concentrations of explosive gases or vapors. Some will indicate warning signals if explosive or ignitable substances are present in the environment. Other meters will indicate concentrations present in contrast to preset lower explosion limits of ignitable gases. It must be noted that personnel should beware of entering ignitable atmospheres in excess of 10% of the lower explosion limit of a given substance.
- Toxic aerosol and gas meters are direct reading devices that warn personnel when hazardous limits of gas or vapor are present. Typical devices would include hydrogen sulfide meters, carbon monoxide indicators, chlorine gas, and others. Some of the more costly devices can be calibrated to detect various levels of several gaseous and aerosol hazards.
- Dosimeters are devices that can be attached to personnel involved in areas containing electromagnetic and ionic radiations. The device is set to indicate a color change or a numerical reading that warns the wearer when high amounts of exposure has occurred.

A positive perspective towards using direct reading passive samplers is that results can be readily obtained. A negative view is that most direct reading instruments require batteries; they need to be routinely recalibrated and routinely maintained. In some cases, a margin of error can be significant.

Active samplers are often used in the realm of industrial hygiene. Personnel may attach "sniffer" devices that pump aerosols through a hose and into a filter over a given amount of time. The filters are periodically changed and taken to a lab for analysis. Analytical results are compared with established OSHA permissible exposure limits for specific airborne hazards. A positive perspective towards using active sampling methods is that reasonably exact results can be achieved. The downside is it sometimes takes significant time to obtain results.

Air Sampling

Some reasons for performing air monitoring are:

- For assessing potential safety and health risks
- To conduct hazards assessments for wearing personal protective equipment
- A means of evaluating environmental pollution potentials
- To aid in selecting actions required for mitigating hazardous conditions
- A mechanism to help determine if decontamination activities were successful

Key physical principles related to air movement, temperature, humidity, and properties of potentially dangerous gases, vapors, dusts, and other aerosols will affect the accuracy of air monitoring. To effectively conduct air monitoring to classify hazards that may affect inhabitants of enclosed spaces, the evaluator should be knowledgeable in the science of psychrometry. This involves comparing air movement, air change, humidity, temperature, substances, people, and other variables. Effective sampling systems or methods, procedures, and equipment chosen to conduct air monitoring depends upon a number of considerations. Figure 4.3 shows factors that should be included in air-sampling considerations.

About Air Monitoring Equipment

Air monitoring equipment should be rugged enough to handle in varieties of potentially hazardous environments. Users of the equipment must be trained to easily use the equipment. Air monitoring equipment should be inherently safe, explosion-proof, and purged from use in previous environmental testing.

Air monitoring equipment must be sensitive, accurate, reliable, and able to provide useful results. The response time, or time required to provide analytical results, should be fast enough to meet objectives for personnel exposed to environmental dangers, especially when using supplied air. In some cases, direct reading instruments may take only seconds, and as much as several minutes, to provide reliable results.

The calibration routine is a key factor for guaranteeing accuracy. Calibration routines are established by the equipment manufacturer and involve using known concentrations of a test chemical to periodically set digital readouts to match the known concentration. Sometimes gases or volatile solvents are used for this purpose (Figure 4.4).

Oxygen and Combustible Gas Indicators (CGIs)

Normal concentrations of oxygen in the air range from 19.5% to 22%. Ambient air should read very close to 20.8%. In conditions involving confined spaces, pits, crawlspaces, and other locations that do not have reasonable air movement, deviations in normal oxygen readings could indicate that other substances are present in the air space. Combustible gas indicators will read lower explosion limits of combustible gases or vapors in correlation to the calibration gas. Please note that CGI's are usually recommended for use in normal oxygen environments only. When atmospheres read in excess of 10% of the lower explosion lever, LEL, (also called lower flammability lever, LFL) personnel should not enter.

Caution: confined spaces should only be entered at 0% LEL.

Toxic Atmosphere Indicators
Colorimetric Indicator Tubes (CIT's) are sometimes used for determining presence of air hazards. They are glass tubes that contain chemical reactants that can indicate the presence of known air contaminants. In order to properly use CIT's, the evaluator must have information about potential contaminants that are likely to be in the air. When used, both ends of the glass tube are broken off and the tube is installed into an air suction device. As air-containing contaminants are pumped through the tube, chemical reactants change color at concentration markings with the tube. Unfortunately, CIT's can produce false readings due to antagonism by other substances also found in the air. If used, the evaluator should consider readings to be qualitative at best.

Photo-Ionization Detectors (PIDs)
Photo-ionization detectors are often used as field quality monitoring devices. They will read non-specific concentrations of organic vapors and gases, as well as some inorganic air contaminants. PIDs use light ionization sensitivity principles that are relative to each substance in order to yield quantitative readings. They have remote sensing capabilities and can yield readings within seconds. They are very sensitive to aromatic and unsaturated hydrocarbon compounds.

Unfortunately, PIDs do not monitor for specific air contaminants and they are not sensitive to hydrogen cyanide or methane. Also, they may not be able to detect some

Considerations for Effective Air Sampling Systems
- The reliability and accuracy for equipment of choice
- Efficiency rating of the equipment and sampling process
- Environmental factors and psychrometric issues
- Portability and ease of use
- Calibration routine
- Equipment serviceability and availability of component parts
- Expectations for information and analytical data

Figure 4.3 Considerations for effective air-sampling systems.

Figure 4.4 Air monitoring equipment.
(Courtesy of the Claycomo, Missouri Fire Department)

halogenated hydrocarbon compounds and they are sensitive to changes in humidity. PIDs are calibrated to detect only one chemical at a time. With these negative considerations, why use a PID? They work well when used as directed by the manufacturer and to detect substances that the equipment senses.

Flame-Ionization Detectors (FIDs)
Flame-ionization detectors detect total hydrocarbon content in the air. They do not analyze specific substances, but will sense total quantities of organic substances contained in the air. They are sensitive to the alkanes or saturated hydrocarbon compounds, and also to the alkenes or unsaturated hydrocarbon compounds. They are portable and use remote sensing probes. FIDs will provide digital readings within seconds.

FIDs are not usable for inorganic air contaminants such as chlorine, hydrogen cyanide, and ammonia. They are less sensitive to some unsaturated hydrocarbon compounds and cannot be used safely in flammable environments. FIDs are less sensitive to organic compounds that contain functional groups such as (-OH) and (-Cl).

Aerosol Monitoring Devices
Aerosol monitoring devices are used to determine the presence of solids and liquids that are contained in the air. They do not directly determine kinds of aerosol contaminants, but do determine aggregate totals. Personnel involved in mold and biohazard evaluations and air monitoring would use this technology.

Radiation Monitoring Devices

Radiation monitoring devices, which include the "old fashioned" Geiger-Mueller counter systems, are used to determine presence of ionizing radiation. Alpha radiation or the least penetrative particle may use proportion counters and scintillation counters. The more penetrative Beta and Gamma radiations use equipment similar to Geiger counters. The topic of radioactivity and monitoring considerations will be covered later in this text.

Materials Sampling

Far too often when company management decides to conduct a "Tidy Friday" cleanup of surplus chemicals, old unmarked chemical containers are found. The object of the cleanup is to decide if surplus materials are taking up needed space or if a need for the substance exists elsewhere in the organization. If a container has been pre-opened, or not recently used, the contents should be questioned. If a barrel is found buried in a field or soil in a vacant lot has a weird smell or color, environmental and personnel safety questions should be answered. Methods used in materials sampling are used for these purposes. The following is a summary of materials commonly sampled and examples of how they may be sampled:

1. Air contaminants - if air is collected into sampling containers and taken to a lab for analysis, materials sampling specifications occur as mentioned in this section.
2. Microbiology samples may be collected from filtered air, obtained from wipe samples taken directly from suspected microbial growth, or transferred to growth media from dust, sticky tape, and other methods.
3. Soil is sampled from surfaces and at measured depths. Chemical and biological sampling is performed using a variety of equipment that may range from a glass jar and tongue depressor to tamper collectors, augers, , and graduated collecting devices.
4. Water can be collected from surface and ground water sources. Surface waters may be collected using pond samplers, bottle samplers, peristaltic pump samplers, and Kemmerer bottles and tubes. Ground waters may be collected using purging methods, gas displacement procedures, submersible pumps, peristaltic pumps, bailers, and other contraptions.
5. Solid chemical sampling usually involves taking a representative portion of the material that is in question that can be compared analytically by volume or weight. Solids may include dusts, grindings, and filings.
6. Liquids, pastes, and sludge chemical sampling is conducted by transferring a sample from a container or holding area by pumping, hydraulic transfer, pneumatic transfer, or penetration.

Note: The potential hazard properties of materials will dictate proper means of sample collection and handling. For instance, ignitable substances would require fire- and explosion-resistant procedures and equipment respectively. Material hazard potentials must be considered in all sampling and handling methods. Personnel involved in sampling potentially dangerous materials must use care to prevent reactions between samples, procedures, and equipment used. Know what you are doing before doing it.

Materials sampling is usually performed when conducting environmental site audits or assessments and mystery materials are found. It is also used for determining if substances are hazardous wastes, otherwise regulated wastes, or to determine if they can be sent to a city landfill. Proper sampling of questionable materials requires following organized procedures or a sampling plan.

Sampling Plan

The prime focus of a sampling plan is to describe methods and procedures for taking a sample from a questionable source and ensuring that the sample remains unchanged before and after chemical analysis. Sampling plans should contain the following elements:

- Background information about the sample such as its source, physical properties, and other known criteria
- Information on how the sample was taken including method, equipment, and procedure. In some cases EPA has required specific, numbered procedures to be used for this purpose.
- Choices of personal protection used by sampling personnel should be included in the sampling plan. In this regard, the sampling plan may possibly serve as OSHA-required PPE hazards assessment documentation.
- Methods used to decontaminate sampling equipment and personnel should be included in the sampling plan.
- The standard operating procedures (SOPs) for preparing to sample, taking the sample, handling, transporting, and analyzing the sample should also be included in the plan.
- Sample integrity should be listed in the plan and included in documents that accompany the sample to the lab. In some cases, certain samples require chemical additives prior to delivering to the lab. Depending upon the analytical method that is used, samples such as wastewater require a specified addition of dilute acid or alkaline.
- To guarantee validity of test results, recordkeeping is required in all cases of materials sampling. Means of recordkeeping should also be described by the sampling plan.
- Proper packaging and shipment of materials test samples should also be described in the sampling plan.

Sample Collection Specifics

In many situations, sampling questionable materials will interface with issues involving regulatory compliance or legal concerns. Because of the delicate nature of how samples are taken, handled, and analyzed, every detail relating to the sampling project must be accounted for. In this regard, the following kinds of information should be noted:

- Site grid layout or mapping - locations where samples were taken are often listed on a drawn grid that shows designated coordinates and references. Photographs are often used for this purpose.
- The number of samples should be noted and referenced to each sampling point listed on the site map.
- Other information would include the weight or volume of each sample taken as well as the type of container used for holding or transferring the sample.

Sample Records, Chain of Custody (Also Called Chain of Evidence)

Irregardless of the kinds of materials that are sampled, chain of custody procedures are required. What is chain of custody? Basically, it's a means of ensuring that a sample remains representative as it was taken from the beginning of sampling to analysis. Typically, verification forms are completed and signed by the sampler, by the transporter, and by the analytical lab. Often testing labs will provide chain of custody forms and containers for environmental samples that they analyze. The following items are commonly included in chain of custody records:

- Description of the sample: microbial, liquid, solid, aerosol
- Weight or volume of material collected
- Sample source: Grid number, well location, barrel number etc.
- Identification of sampler
- Date and time sample was taken
- Weather conditions (if applicable)
- Temperature specifications (keep cold, warm or ambient)
- Sample tracking number
- Method(s) used to take samples
- Chemical additives or temperatures used for preservation
- Container type
- Method used to seal container and person who verified the seal
- Reason for sampling
- Sample appearance: color, turbidity, sludge, solid
- Regulatory information: EPA or OSHA guidance references
- Name and address of organization preparing and transporting sample

- Name and address of organization that analyzes the sample
- Date and time sample was analyzed
- Results determined from analysis
- Signature and certification of analysis

RISK ASSESSMENT

So far in this chapter we have considered methods for evaluating site hazards and environmental contamination monitoring, sampling, and analysis. We have discussed methods, procedures, and equipment that are used in conducting site evaluations. At this time, we want to use information that has already been covered and blend it together with other considerations that need to be evaluated before entering potentially hazardous conditions.

Whether an environmental health and safety professional is evaluating an industrial worksite or is planning to enter an unknown environment, key considerations need to be addressed. Understanding potential risks at a given site helps us to ensure that we leave the site with the same health conditions as we entered. How we conduct operations, perform environmental monitoring, or conduct work activities within potentially hazardous environments requires us to consider the following before committing resources.

At no time should personnel be allowed to perform alone in hazardous environments—The buddy system prevails.

Elements of Risk Assessment

Health and safety potentials must be considered, based upon known properties of potentially dangerous substances on-site. Methods, procedures, and personal protection are based upon this portion of the site assessment.

Understanding volatile conditions that may cause fire or explosions to occur is important. How reactive substances are stored and used on site must be evaluated.

For personnel that may be involved in working within confined spaces and locations with poor ventilation, oxygen levels must be evaluated and managed. As previously mentioned, normal ambient oxygen exists at levels between 19.5 % and 20.8 %.

- If oxygen content in air is less than 19.5%, air-purified respiratory protection cannot be used. Supplied air using self-contained breathing apparatus or air line supply is required.
- If oxygen content in air is less than 16% and air is not supplied, entry personnel will be subjected to anoxia conditions that can easily lead to death.

Oxygen-rich environments are dangerous. Is there such a thing as too much oxygen? Yes. Remember that oxygen is not flammable; it makes everything else flammable or explosive. Personnel should be aware if oxygen content in air can exceed 22%.

Be aware of biological hazards. In poorly ventilated spaces, confined spaces, locations containing moisture, or conditions that may yield microbiological hazards, don't enter without further evaluating the space. Bugs, poisonous plants, and other unfriendly living critters need to be evaluated before entering their environment. Sometimes proper personal protective equipment handles this consideration, but not in all cases.

When personnel enter a site after a fire, storm, flood, structural collapse, or when other work activities are occurring, electrical hazards are usually present. Open or exposed wiring in damp environments present exceptionally dangerous conditions for site entry personnel. Conditions relating to chemical reactivity, explosive and flammable atmospheres are more hazardous when electrical hazards exist. Uncontrolled or non-locked and tagged hazardous energies from sources that include electrical, pneumatic, pressurized water and steam, mechanical, and other sources increase dangers in hazardous environments.

Special provisions are required for personnel expected to perform in hot and cold environments. After donning total coverage personal protective equipment personnel are exposed to conditions that can yield heat stress and if left unchecked, heat stroke. An important element of personal protective equipment training includes listening to your body and being aware of early symptoms of heat stress. Cold temperatures also add to problems related to movement, dexterity, and potentials for frost bite.

When working in noisy environments, personnel are required to wear hearing protection or the source of noise must be reduced. When noise exceeds OSHA/NIOSH permissible exposure limits, employees must be placed into a regulated hearing conservation program. In these conditions damage can be caused to the inner ear that results in hearing impairment and hearing loss. Also, problems in communication and mental or emotional concerns may arise, especially when noise is coupled with other hazardous activities.

Personnel may be exposed to varieties of electromagnetic radiation hazards in the course of performing their duties. In this regard personnel should also be aware of conditions caused by non-ionizing (NI) radiation hazards. Sources of NI radiation include the sun and other light-related hazards. The basic effect of NI radiation upon the body is skin damage (cancer) and eye damage (retinal), however others may be of issue in coming years. Ionizing radiation involves emission of penetrative radioactive energies from radioactive materials and includes alpha, beta, and gamma radiations. Ionizing radiations cause varieties of internal and external body damage including burns, cellular mutation, organ damage, cancer, and others. Awareness, protection, and safe distance are stressed when working with ionizing radiation. Physical safety hazards, including confined spaces, inherent site hazards, and other considerations are covered in the hazards' assessment checklists that follow.

ENVIRONMENTAL HEALTH AND SAFETY
Site Hazards Evaluation Checklist

Location _____ Area_____ Date_____

Responsible Person_____

Evaluated By_____ SIC _____

Description of Site Function _____

ENVIRONMENTAL PROGRAMS
(A). HAZARDOUS/REGULATED MATERIALS

1. Does facility store chemicals?	Y	N	NA
2. Are materials usage inventories available?	Y	N	NA
3. Regulated materials used on-site Manufactured Processed Otherwise Used			
4. Are extremely hazardous materials stored?	Y	N	NA
5. Are storage inventories of HAZMAT Available?	Y	N	NA
6. Are regulated materials stored in tanks?	Y	N	NA
7. Are VOC chemicals used?	Y	N	NA
8. Are HAP chemicals used?	Y	N	NA
9. Are Persistent Bio-accumulative Toxics (PBT) used	Y	N	NA
10. Are regulated materials emitted from this site?	Y	N	NA

(A) NOTES

1_____

2_____

3_____

4_____

5_____

6_____

7_____

8_____

9_____

10_____

(B). HAZARDOUS /REGULATED WASTES

1. Are hazardous wastes generated from this site?	Y	N	NA
2. Does site have a written waste management plan?	Y	N	NA
3. Waste generator status	Large	Small	Conditionally Exempt
4. Are medical or biohazard wastes generated?	Y	N	NA
5. Is site state and federal registered?	Y	N	NA
6. Does site generate universal wastes?	Y	N	NA
7. Does site store wastes?	Y	N	NA
8. Does site recycle wastes?	Y	N	NA
9. Does site generate rags for cleaning?	Y	N	NA
10. Have employees received RCRA Training?	Y	N	NA

(B) NOTES

1 _____

2 _____

3 _____

4 _____

5 _____

6 _____

7 _____

8 _____

9 _____

10 _____

(C). WASTEWATER

1. Do on-site processes generate wastewater?	Y	N	NA
2. Does site have a wastewater permit?	Y	N	NA
3. Does site have a storm water permit?	Y	N	NA
4. Is wastewater drained into storm drains?	Y	N	NA
5. Are floor drains near chemical storage areas?	Y	N	NA
6. Are regulated materials disposed in wastewaters?	Y	N	NA
7. Does site treat wastewaters?	Y	N	NA

(C) NOTES

1_____

2_____

3_____

4_____

5_____

6_____

7_____

(D). AIR

1. Are fume hoods used on site?	Y	N	NA
2. Are heated processes used on site?	Y	N	NA
3. Are vents routinely inspected?	Y	N	NA
4. Does site have an air permit?	Y	N	NA
5. Are air-regulated materials released on site?	Y	N	NA
6. Are emissive processes used on site?	Y	N	NA

(D) NOTES

1_____

2_____

3_____

4_____

5_____

6_____

(E). ENVIRONMENTAL REPORTING

Has site submitted any of the following environmental reports for any year?

1. Tier 1 EPCRA, MSDS or Chemical Listings	Y	N	NA
2. Tier II EPCRA Hazardous Materials Storage	Y	N	NA
3. Air Emission Inventory EIQ	Y	N	NA
4. Toxic Release Inventory Form R	Y	N	NA
5. Quarterly Hazardous Waste Reports	Y	N	NA
6. Biennial Hazardous Waste Reports	Y	N	NA
7. City Waste Water Reports	Y	N	NA
8. City Air Reports	Y	N	NA
9. Others	Y	N	NA

(E) NOTES

1_____

2_____

3_____

4_____

5_____

6_____

7_____

8_____

9_____

HAZARDOUS MATERIALS COMMUNICATION

Site Evaluation Checklist

Location _____ Area_____

Responsible Party_____ Date _____

Evaluated by _____

Management Review by_____

A. Overall Site Layout

1.	Lighting	Y	N	NA
2.	Location of flammable or explosive substances	Y	N	NA
3.	Entry, exit and passages open	Y	N	NA
4.	Damp or wet locations	Y	N	NA
5.	Proper storage cabinets & shelves	Y	N	NA
6.	Blocked utility access	Y	N	NA
7.	Exits			
	a. Illuminated	Y	N	NA
	b. Obstructions	Y	N	NA
	c. Secondary egress routes available	Y	N	NA
	d. Fire exit doors operable	Y	N	NA
	e. Emergency exits open	Y	N	NA
8.	Trip or fall hazards	Y	N	NA

B. Emergency Planning

Operations

1.	Fire extinguishers on site	Y	N	NA
2.	Fire extinguishers readily accessible	Y	N	NA
3.	Safety showers and eyewash showers near by	Y	N	NA
4.	Emergency alarm system on site	Y	N	NA
5.	Emergency lighting operable	Y	N	NA

C. Plans, Procedures, and Placards

Information

1.	Site emergency action plan available?	Y	N	NA
2.	MSDS copies readily available?	Y	N	NA

3. Is Site hazard communication plan written and up-to-date?	Y	N	NA
4. Is site respiratory protection plan written and up-to-date?	Y	N	NA
5. Is personal protective equipment plan written, contains assessment and up-to-date?	Y	N	NA

Placards & Markings

6. Site evacuation placards posted	Y	N	NA
7. NFPA 704 placards posted	Y	N	NA
8. Containers, pipes, and vessels marked	Y	N	NA

D. Personal Protective Equipment

1. Full body coverage PPE available A or B	Y	N	NA
a. Protection Levels C	Y	N	NA
b. Protection Levels D	Y	N	NA
2. Positive pressure SCBA required	Y	N	NA
3. Air-purified respiratory protection required	Y	N	NA
3. Respirator users:			
a. Routine maintenance & proper storage	Y	N	NA
b. Inspection documentation available	Y	N	NA

E. Chemical Storage

Storage Locations

1. Safe storage in open areas	Y	N	NA
2. Cabinets and storage buildings	Y	N	NA

Containers

3. OSHA & DOT markings	Y	N	NA
4. Closed when not in use	Y	N	NA
5. Integrity of containers	Y	N	NA

Logistics

6. Compatible materials stored together	Y	N	NA
7. Floor or storm drains nearby	Y	N	NA
8. Drums, boxes or pallets stored outside	Y	N	NA
9. Proper handling and moving equipment	Y	N	NA

F. Flammable & Combustible Liquids

1. Ventilated when used	Y	N	NA
2. Minimal open containers in a fire regulated room	Y	N	NA
3. Explosion and fireproofing, layout and equipment used	Y	N	NA

G. Compressed Gases

1. Properly stowed	Y	N	NA
2. Away from traffic	Y	N	NA
3. Empty storage	Y	N	NA
4. Markings	Y	N	NA
5. Proper means of transporting	Y	N	NA

H. Waste Disposal

1. Regulated waste management checklists being followed	Y	N	NA
2. Labeled with RCRA & DOT markings	Y	N	NA
3. Container tracking system being used	Y	N	NA
4. Storage per RCRA and other regulatory directives	Y	N	NA

I. Employee Training

Training

1. OSHA hazard communication training conducted	Y	N	NA
2. Training documented	Y	N	NA
3. Emergency action & RCRA training conducted	Y	N	NA
4. Training requirements documented on job descriptions	Y	N	NA
5. Trainers Certified	Y	N	NA

J. Employee Review

1. Employees knowledgeable about HAZCOMM Program	Y	N	NA
2. Accessible to MSDS system	Y	N	NA
3. Understand HAZCOMM, DOT & RCRA labels	Y	N	NA
4. Understand how to choose PPE and how to prevent exposure to hazardous substances	Y	N	NA

OBSERVATIONS

RESPIRATORY PROTECTION

Hazards Evaluation Checklist for APR Respirator

Location _____ Area _____

Responsible Party _____ Date _____

Evaluated by _____

Management Review by_____

A. Respiratory Protection Program

1. Written respirator plan, up-to-date   OK   NO   NA
2. Affected employees receive required training (annual)   OK   NO   NA
3. Training and fit test documented   OK   NO   NA
4. Affected employee medical evaluation documented   OK   NO   NA

B. Inspection for APR

1. Condition of face piece (clean, damaged, serviceable)   OK   NO   NA
2. Lens (clean, damaged, serviceable)   OK   NO   NA
3. Adjustment straps (worn, serviceable)   OK   NO   NA
4. Attachments (worn, serviceable)   OK   NO   NA
5. Valves serviceable (inhalation, exhalation)   OK   NO   NA
6. Single direction, check valves (worn, serviceable)   OK   NO   NA
7. Attachments, valve covers, and fittings (worn, serviceable) OK   NO   NA
8. Filter attachments, gaskets etc.. (serviceable, worn)   OK   NO   NA
9. Proper filters used for specific hazard   OK   NO   NA
10. End of life indicator system being used   OK   NO   NA
11. Overall respirator condition (clean, serviceable)   OK   NO   NA
12. Respiratory storage (bagged, container)   OK   NO   NA
13. Respiratory inspection documents available   OK   NO   NA

OBSERVATIONS

_____

_____

_____

_____

_____

CONFINED SPACE ENTRY
Hazard Evaluation Checklist

Location _____ Area_____ Date _____

Responsible Party _____

Evaluated by _____

CONSIDERATIONS

A. Site Confined Space Evaluation

    1. Confined spaces monitored on-site   OK   NO   NA

    2. Any designated as permit spaces (placards)   OK   NO   NA

    3. Written confined space entry program (up-to-date)   OK   NO   NA

    4. Provisions in place for confined space rescue   OK   NO   NA

    5. Training for confined space & rescue personnel   OK   NO   NA

    6. Proper equipment on site   OK   NO   NA

B. Confined Space Entry Procedures/Entry Team Communication

    1. Procedures in place for pre-entry evaluations   OK   NO   NA

    2. Permits completed, protective measures defined   OK   NO   NA

    3. Tasks defined, safety and entry equipment in place OK   NO   NA

    4. Entry personnel equipped with PPE, harnesses   OK   NO   NA

    5. Ventilation needs to be defined and in place   OK   NO   NA

    6. Air and hazards monitoring in place   OK   NO   NA

    7. Hazardous energy control (lock out/tag out) in place OK   NO   NA

    8. Entry and rescue team assignments made   OK   NO   NA

    9. End of shift communication plan established   OK   NO   NA

    10. Documentation of completed confined entry operation   OK   NO   NA

OBSERVATIONS

_____

_____

_____

_____

_____

_____

VENTILATION
Hazards Evaluation Checklist

Location _____ Date_____

Responsible Party_____ Evaluated by _____

Management Review by _____

CONSIDERATION

Chemicals, MSDS evaluated for venting requirements at emission sources?	Y	N	NA
HVAC system routinely inspected?	Y	N	NA
Are fume hoods checked for proper operation?	Y	N	NA
HVAC system routinely inspected for holes, leaks, corrosion	Y	N	NA
Air velocities routinely checked at entry ports?	Y	N	NA
Inspection velocities dated & labeled at each ventilation entry port?	Y	N	NA
Entry ports near contamination sources?	Y	N	NA
Is cross ventilation required at contamination emission sources?	Y	N	NA
Is make up air proportionate to exhaust venting needs?	Y	N	NA
Exhaust air flow balanced between all ports?	Y	N	NA

OBSERVATIONS

_____

_____

_____

_____

_____

_____

Chapter 5

HAZARDOUS MATERIALS EMERGENCIES

THE BEST TIME TO PLAN FOR AN EMERGENCY IS BEFORE ONE OCCURS

Within this text we have declared that hazardous or dangerous materials are substances that can harm people or the environment. Hazardous wastes are the same substances that no longer have value. Stuff happens that changes useful materials into wastes, often within a wink of the eye. Sometimes non-dangerous materials become dangerous as the result of fire or some other physical reaction. At other times, incidents occur that yield releases of dangerous substances into our neighborhoods and ecosystems and they may remain in our immediate environment.

Unfortunately, there are also incidents when some idiot decides to prove a point by subjecting innocent people to dangerous conditions caused by hazardous reactions or by exposing them to chemical, biological, or radiation hazards. The limits of these unwarranted actions are controlled only as a deviate mind may consider.

Industries and special interests from the past, as well as powerful people in today's world that lack consideration, have presented the people of this world with environmental and human health conditions that are likely to take several lifetimes to remedy. As the Vietnam War ended, it became apparent that many places across the globe were spotted by man made hazardous conditions that rival only Pandora's Box.

Some hazardous environmental conditions occurred through the advent of rising technologies, rapidly growing industries, as well as evil doings that infected our world by actions of warring nations and unscrupulous businesses. In recent years, it has been manifested that most of the major nations of the world, as well as mid-eastern and European nations that hate them, harbor weapons of mass destruction. Materials were produced, warehoused, and hidden and are so dangerous that they can destroy our earth, as well as several other planets if they are released.

Figure 5.1 Chemical emergency team members never respond alone. *(Courtesy of Joseph Hilliard and HMTRI/CCCHST)*

Is there a rising need for hazardous waste operations and emergency response personnel? Everyday incidents require immediate action and response by properly trained and properly equipped personnel. Also as the crypts are opened, the world should only hope that committed, properly trained and properly equipped individuals are available to fight, conquer, and remove sleeping dragons that were born yesteryear.

ABOUT HAZWOPER

We mad chemists and engineers were once responsible to handle any chemical emergency that occurred in our workplaces. After all, we usually had something to do with processes or materials involved in the emergency. We only needed to find someone crazy enough to put on gear and follow us into action. It always is a wonder why principles of "away from the lab" dangerous materials safety was not part of the collegiate curriculum. At one time EPA called all shots for actions conducted to deal with contaminated communities and to clean up hazardous waste sites. Even though OSHA provides safety and health guidance for all American workers, EPA was the coordinating authority. Many law enforcement personnel and firemen have entered situations without knowledge of the hazards they would be facing and without proper equipment, training, and direction to deal with dangerous materials. Why? In many cases this stemmed from city politics and lack of funding. Even today this is an issue for outlying communities.

During the 1980's the Occupational Safety and Health Administration received responsibility to develop the Hazardous Waste Operations and Emergency Response (HAZWOPER) Standard. The standard went into effect in 1990. About the same time OSHA insisted that chemical manufacturers provide material safety data sheets (MSDS) to users of hazardous products. Also EPA, state emergency response commissions, and local emergency planning committees fostered and further developed programs driven by the emergency planning and the community's right to know.

The following personnel are required to be covered by the HAZWOPER emergency standard:

- Any worker required to respond emergencies involving dangerous substances
- Any worker required to clean up and manage chemical or biohazard spills
- Hazardous waste site workers
- Chemical abatement contractors
- Biohazard abatement contractors
- Disaster and waste site decontamination personnel
- Emergency medical personnel
- Industrial emergency response and HAZMAT team personnel
- Fire department HAZMAT Team personnel
- Fire fighters with hazardous materials responsibilities
- Law enforcement personnel with hazardous materials responsibilities

Following are the needs of HAZWOPER personnel:

- They need to be trained to do whatever they are expected to do.
- They need to be provided support, direction, equipment, and all needed resources, including funding.

For direction in managing HAZWOPER programs, refer to the OSHA Hazardous Waste Operations Standard, 29 CFR 1910.120 for general industry and 29 CFR 1926.65 for construction. For training direction refer to Appendix E of the HAZWOPER Standard and for direction in training quality, refer to the ANSI Z490 Standard, Accepted Practices in Safety, Health and Environmental Training.

EMERGENCY PLANNING

What is a hazardous materials emergency? In short, a HAZMAT emergency is an incident that requires immediate action to stabilize, control, mitigate, and remedy the cause and results of uncontrolled spills or releases of hazardous substances. The following may be considered hazardous materials incidents:

- Explosion or fire: fumes, gases, particles, and fire reaction by products
- Environmental abnormalities: odors, vapors, fumes, smoke, deterioration of containers and equipment
- Incompatibilities: swollen containers, fuming and gassing, formation of hazardous residues
- Biological hazards related to spills and releases of human body fluids; also other health hazards that may involve human and animal fecal materials, decomposition of organic materials, and endemic transfers of microbial pathogens
- Spills and releases of hazardous or otherwise regulated substances
- Uncontrolled releases of chemicals or biohazards
- Natural disasters that release or form dangerous materials from waterborne sources such as floods of ground water, sewers, and seawater. This includes flooding by clean (Type A.1), gray (Type A.2), and black (Type A.3) classifications of water that are subjected to organic materials, extended periods of time, and bioactive temperatures that support microbial production of pathogenic and toxic materials.
- Acts of chemical, biological, and radiological terrorism

Contingency

Contingency should answer the question, "What are you going to do when stuff happens in a manner that you didn't expect it to?" The most important part of responding to any hazardous materials emergency is the initial evaluation or assessment of what has occurred and decisions that were predicated upon the initial evaluation. In all cases, the buddy system prevails. Personnel shall never respond alone to hazardous materials emergencies.

Elements of Incident Response

Depending upon the site, available resources and abilities of site personnel respond-ing to hazardous materials incidents may only involve personnel evacuation. Some locations may require partial to full-scale aggressive actions taken by on-site person-nel. Irregardless of actions taken, all responses to hazardous material emergencies require following site-specific emergency action plans and procedures. These ele-ments are discussed in the following sections.

Correlation Between Emergency Action and Emergency Response Plans
Every place of employment is required to have a written and functional OSHA-required emergency action (EA) plan. In some circles this plan is called the fire and evacuation plan. The prime focus of the EA plan is to describe actions that employ-

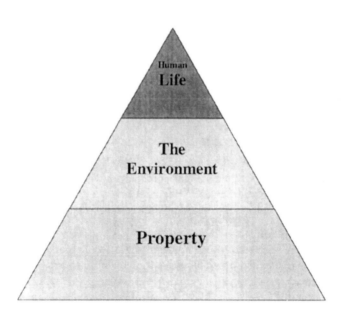

Figure 5.2 Emergency response triad of importance.

ees are expected to take at times of emergency. If any employee is required to perform a specific emergency duty, the emergency action plan describes specific duties
and procedures that shall be followed. If emergencies are determined to be hazardous materials emergencies, a written and functional emergency response (ER)
plan is required to dovetail or correlate with the site emergency action plan. During
site audits, the first documents that most safety and health inspectors ask to review
are the emergency action and emergency response plans. Samples of emergency
action and emergency response plans are available in this text.

Emergency Response (ER) Plans
Also called hazardous materials contingency plans, emergency response plans are
required to describe all aspects of responding to hazardous materials emergencies at
any given site. Limitations of response are to be delineated and made clear within
the plan. ER plans should emphasize the triad of importance as shown in Figure 5.2.

The Emergency Response Triad of Importance
Following are the priorities of importance for the emergency response triad and their
explanations:

- The first priority of emergency response actions shall focus upon preserving
  human life at the incident site and in surrounding communities.
- The second priority is to contain, if possible, all spilled and released dangerous
  substances and to make every effort to prevent hazardous emissions from entering the air, waters, or the land.
- The third priority is to prevent damage to property, structures, and equipment.

Emergency response plans are to be implemented for the purpose of controlling
anticipated hazardous material emergencies prior to commencing with any emergency response operation. The plan should be written and available for review by
employees, their representatives, and by regulatory inspection personnel. Emergency
response plans shall:

- Be written as official plans
- Contain pre-emergency planning and coordination with organizations from the
  community and other resources
- Define response and support personnel roles, lines of authority, training, and
  communication
- Describe methods of emergency recognition and prevention
- Describe methods and procedures used to conduct hazards evaluation and
  assessment
- Describe applications for chemical protective clothing and personal protective
  equipment used for responding to hazardous materials incidents that may occur
  on site

- Provide direction in locating or obtaining emergency equipment to be used for responding to emergencies on site
- Emphasize that personnel shall not respond alone, the buddy system prevails
- Describe locations of safe distance and refuge
- Prescribe measures to be taken to ensure site security and control
- Clearly define routes of egress and evacuation
- Prescribe methods and procedures used for mitigating emergency conditions and for controlling the cause of an emergency
- Explain methods and procedures used for site decontamination
- Describe actions to be taken for emergency medical treatment and for conducting first aid
- Provide guidance in emergency alerting
- Provide guidance in conducting overall response procedures

Site-Specific Emergency Response Plans
Site-specific emergency response plans are usually written to direct actions for the following hazardous materials incident response scenarios:

- To provide direction for safe evacuation of employees and to summon help from external emergency response agencies and resources. Note 1: Organizations that direct personnel evacuation to safe distances only, to call for outside support, and do not expect personnel to perform any emergency actions may only be required to implement an emergency action plan. Note 2: Per OSHA, "Incidental releases of hazardous substances that can be absorbed, neutralized, or otherwise controlled at the time of release by employees in the immediate area or by maintenance personnel are not considered emergency responses."
- To provide direction for employees that partially respond to hazardous materials emergencies only for the purpose of controlling hazardous spills and releases from safe distances until off-site emergency responders arrive on scene and to provide support for external HAZMAT response agencies as required.
- To provide direction to employees that aggressively respond to hazardous materials emergencies to control, stabilize, and correct sources of spills and releases and to provide contingency guidance if conditions occur that are beyond the abilities of on-site response personnel.

EMERGENCY PERSONNEL

On-site personnel that respond to hazardous materials emergencies shall be trained to follow directions given in the site-specific emergency response plan. Off-site fire department HAZMAT teams and other HAZWOPER-affected personnel shall be trained under emergency response plans that may encompass many varieties of emergency response procedures.

On-Site Personnel

Emergency response plans identify all individuals and teams that are expected to respond to hazardous materials emergencies that occur on site. According to OSHA, "an emergency response team is an organized group of employees, designated by the employer, who are expected to perform work to handle and control actual or potential leaks or spills of hazardous substances requiring possible close approach to the substance. Team members perform responses to releases or potential releases of hazardous substances for the purpose of control or stabilization of the incident. A HAZMAT team is not a fire brigade nor is a typical fire brigade a HAZMAT team...However, a HAZMAT team may be a separate component of a fire brigade or fire department" (reference: 29CFR 1910.120 [a][3]).

On-Site Emergency Response Team Leadership

When an emergency occurs, one person should take charge and several other team members should work with the team leader to manage the incident. All team members should be trained to function as one unit that understands individual responsibilities and works together to properly stabilize emergency conditions as directed by the team leader, also called Incident Commander or Senior Response Official.

Incident Commander (IC) or Senior Response Official (SRO)
The Incident Commander or Senior Response Official must be empowered to assume control of hazardous materials incidents from the beginning until the incident is released and the work area returns to service. Whenever outside support is requested and the fire department emergency response team arrives on scene, the on-site SRO will brief the HAZMAT Team IC and offer support when requested by the responding team. The following example is typical of incident response actions that on-site Senior Response Officials (SRO) are responsible to manage. Please note that the order listed does not necessarily indicate actual sequencing of events. See Figure 5.3 for a graphic representation of HAZMAT incident response.

What happens when an emergency occurs? Following is the sequence of events:

1. A hazardous materials incident occurs.
2. Site Security is notified.
3. Site Security notifies the on-site SRO.
4. The Senior Response Official contacts team members.
5. The response team meets the SRO at an assembly point.

6. SRO sizes up and assesses the incident with the team and resources.
7. Response actions and equipment needs are determined after variables and resources are further assessed and evaluated.
8. Response zone perimeters are established.
9. The response operations (OPS) leader is assigned and briefed by SRO.
10. The response Safety and Health Officer is assigned and briefed by SRO.
11. The Incident Command Center is established.
12. The OPS control center is established.

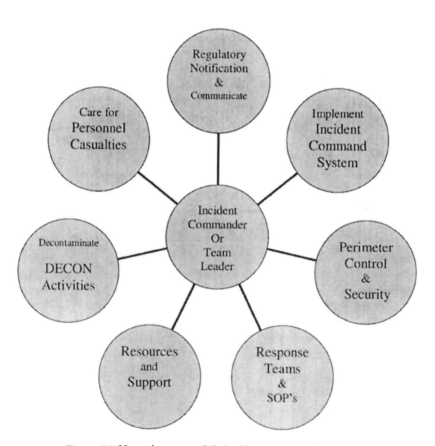

Figure 5.3 Hazardous materials incident command decisions.

13. Perimeter control assignments are made by OPS.

14. Notifications of personnel casualties are made to emergency medical service as directed by the SRO.

15. If casualties and personnel injuries are involved, victim recovery and first aid teams are assigned, briefed, and respond as needed.

16. Notifications to regulatory agencies are made for reporting releases to the environment, as directed by the SRO.

17. The first response entry team is assigned, briefed, and monitored by OPS.

18. Incident entry support and back up team is assigned and briefed.

19. Decontamination team is assigned and DECON corridor is assembled.

20. Mop up and closure team is assembled, briefed, and evaluates incident area.

21. Response results are evaluated and incident area is returned to service.

As shown in Figure 5.3, actions performed and decisions made by the Incident Commander or Senior Response Official usually occur as a sunburst of activity, unfortunately, not usually enumerated. These are some of the items that a response team leader must consider, often all at once. In reality, this matrix will grow as incident variables grow.

INCIDENT CLOSURE

After all emergency response actions are completed and incident areas have been tested and determined safe for normal occupancy, the incident area is released. At this time the Senior Response Official meets with all responders and ensures that the following events occur:

- Response personnel are debriefed.
- Response actions, methods, and procedures are critiqued.
- SRO and responders discuss potentials for improving future emergency action events.
- The sequence of events and response actions that were performed are documented.
- Action items requiring additional work and corrections are assigned for follow up, and a follow-up timeline is established.
- Applicable written reports are provided internally to site management and to site management designees.
- Applicable written reports are provided off-site to regulatory authorities and others as required.

ABOUT EMERGENCY RESPONDERS: ON-SITE EMERGENCY TEAM TRAINING

The value of on-site emergency response teams is that they are comprised of personnel already located on the site. All emergency response personnel are required to demonstrate competency in proper emergency response actions before being assigned to respond. Also refer to the HAZWOPER Standard, Appendix E. Key elements of training that on-site emergency responders are required to have include:

- Understanding the site emergency action and emergency response plans
- Knowing how to implement site emergency response plans
- Knowing how to implement the incident command system
- Familiarity with the site layout
- Know locations of all hazards and hazardous materials inventories on site
- Understand processes and equipment used on-site
- Knowing how to safely deal with the hazardous properties of dangerous materials and processes used on-site
- Having been medically evaluated for use of respiratory protection and other forms of personal protective equipment
- Having been properly fitted with respiratory protection and other forms of personal protective equipment
- Being competent in using chemical protective equipment and all other forms of personal protective equipment
- Know where all emergency equipment is located on site
- Being competent in using all emergency equipment relative to potential emergencies that may occur on site
- Understanding how to follow safe operating procedures especially under emergency conditions
- Being competent in following specific emergency response standard operating procedures
- Being competent in first aid and decontamination of injured personnel
- Being aware of individual and team response limitations
- Understand how to activate contingencies with off-site emergency response agencies and know when to implement them
- Knowing how to interface with off-site emergency support personnel

COMMUNITY EMERGENCY TEAMS

Personnel that may respond to emergencies at a variety of incidents, on private properties as well as within the community, must be equipped and proficient in all the competencies listed above and also be extensively trained in dealing with unknown hazard potentials. In this regard, communities that choose to provide fire department or other types of community HAZMAT team support should have a full time HAZMAT team program. Because of vast differences in potential incident areas, increased dangers, and increased potential liabilities, community-sponsored emergency hazardous materials responders should only have hazardous materials responsibilities.

CLASSIFICATION AND DUTIES OF EMERGENCY RESPONSE PERSONNEL

As described in the OSHA Hazardous Waste Operations and Emergency Response Standard, emergency response personnel shall be assigned duties only after obtaining the skills and knowledge that are required to safely participate as an emergency response team member.

The level of participation for each team member is based upon performance expectations, training, and demonstrated competencies. The following are examples of required first responder skill levels.

First Responder Awareness Level

Personnel at this level are likely to witness or discover incidents that involve hazardous materials. Awareness level responders must know how to initiate appropriate emergency actions by notifying proper authorities about the incident. They would not take any further action to stabilize emergency conditions.

First Responder Operations Level

Personnel at this level would respond to spills, releases, and potential releases of dangerous substances. They would take part in the initial response for the purpose of protecting nearby people, property, or the environment from effects of the incident. They perform at a defensive level without trying to stop the cause of the incident. Operations level responders would take action from a safe distance to prevent spreading of released dangerous materials and to prevent hazardous exposures on site as well as to the surrounding community. Operations level responders should have sufficient experience to objectively demonstrate emergency response competency and should have at least eight hours of formal training at the operations level. At minimum their knowledge should be certified by the employer to include:

a. Basic hazard and risk assessment techniques
b. Selection and use of chemical protective clothing and appropriate personal protective equipment used for operational level actions

c. Basic knowledge about properties of dangerous substances that they may encounter

d. Knowledge in hazardous substance control, containment, and/or confinement activities as related to any site that they may respond to

e. How to implement and perform decontamination procedures

f. Knowledge of relevant hazardous materials response standard operating procedures

g. Knowledge in implementing incident termination procedures

First Responder Technician Level

Personnel at this level would respond to spills, releases, and potential releases of dangerous substances for the purpose of stopping the cause of the incident. Technician level responders take an aggressive role in order to stop the source of hazardous materials spillage or release. Employers must certify that technician level responders have successfully completed 24 hours formal training equal to the operations level and that they have demonstrated technician level competencies that include:

a. Knowing how to implement the site emergency response plan

b. Knowing how to classify, identify, and verify known and unknown substances by using field survey instruments and equipment

c. Understanding the Incident Command System (ICS) and knowing how to function within the ICS

d. Knowing how to select and use appropriate, specialized personal protective equipment to perform technician level activities

e. Understanding hazard evaluation and risk assessment techniques

f. Knowing how to implement and perform decontamination procedures

g. Understanding how to implement and perform incident termination procedures

h. Understanding the properties of dangerous materials and knowing how to predict their behavior

First Responder Specialist Level

Personnel that respond at this level will provide support to hazardous materials technicians. Their duties are focused upon knowledge of various dangerous substances and specifics related to each. The First Responder Specialist would also interface with federal, state, local, and other governmental authorities regarding incident response activities. Employers shall certify that Specialist Level Responders have successfully completed at least 24 hours training at the technician level and that they have demonstrated specialist level competencies that include:

a. Knowing how to implement the site emergency response plan

b. Understanding classification, identification, and verification of known and unknown substances through the use of advanced survey instruments and equipment

c. Understanding the state emergency response plan

d. Knowing how to select and use appropriate specialize chemical protective clothing and personal protective equipment used at the technician level

e. Understanding in-depth hazard and risk assessment techniques

f. Being able to perform specialize control, containment, and or confinement operations that are within capabilities of resources and personal protection capabilities

g. Knowing how to determine and implement methods and procedures for decontaminating affected incident locations

h. Be able to develop a site safety and control plan

i. Knowing the properties and specifics related to controlling dangerous substance releases and having the ability to predict their behavior

Incident Commander or Senior Response Official

The IC or SRO will assume control of the incident scene beyond the first responder awareness level and shall have successfully completed training at least equal to the first responder operations level and shall have demonstrated incident command competencies that include:

a. Being able to implement and operate under the incident command system

b. Knowing how to implement the site emergency response plan

c. Understanding hazards and risks that may exist on site

d. Understanding hazards and risks for employees that enter the site

e. Understanding hazards and risks associated with employees using personal protective and other related equipment

f. Knowing how to implement appropriate community, city, state, and federal emergency response plans

g. Understanding the operation and potentials for support by state and or federal regional response teams

h. Knowing the importance and methods of using appropriate decontamination procedures

i. Understanding how to properly control emergencies and terminate the incident and incident command procedures

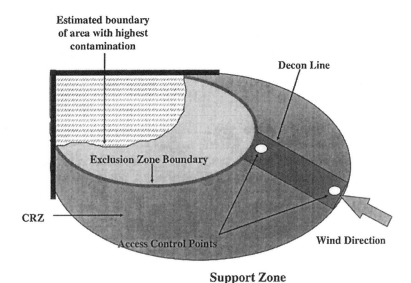

Figure 5.4 Incident control—Hazardous materials incidents are segmented into 3 primary control zones (HMTRI).

INCIDENT MANAGEMENT

The prime focus of responding to a hazardous materials emergency is to:

1. Stabilize the emergency
2. Control the incident
3. Mitigate the results of the incident
4. Correct the cause of the incident
5. Decontaminate the incident site
6. Return the incident site back to normal operations

As shown in Figure 5.4, the incident site must be controlled immediately. All necessary personnel must evacuate and the perimeters must be established in order to safely perform incident management procedures and to control entry and allow authorized personnel to enter and exit the site. Incident management perimeters are established in three (3) primary zones:

1. The Hot Zone or Exclusionary Zone: This is the place that the incident occurred.

2. The Contamination Reduction or DECON Zone: This is the place that entry personnel, casualties, and equipment are decontaminated.

3. The Support Zone: This is a reduced hazard location where incident support functions are staged and a controlled location where incident management, incident entry, and backup personnel may rest, be briefed, as well as supporting other non-hazardous activities.

Note: See in Figure 5.4 that there is only one way into the incident area and only one way out. Hazardous materials incidents are segmented into 3 primary control zones: exclusionary zone or contamination zone; contamination reduction zone or decontamination corridor; support zone.

Note that site entry and exit is controlled from one point, through the decontamination corridor that is located upwind of the incident.

Figure 5.5 Hazardous materials incident command system.

INCIDENT COMMANDER

As mentioned before, only one person should be in charge when a hazardous materials incident occurs. Please note in Figure 5.5, even though only one person is in charge of a hazardous materials incident, that one person is advised and supported by several other people and resources. The Incident Commander or Senior Response Official that manages a hazardous materials incident must be knowledgeable in incident command procedures. It is highly recommended that IC or SRO personnel should complete a formal course in incident command systems. There are both on-line computer courses and in-class courses in Incident Command Systems that are available from the Federal Emergency Management Agency, National Emergency Training Center at Emmetsburg Maryland. Figure 5.5 is an example of a typical incident command system used to manage hazardous materials incidents. This is an example of a command structure that an incident may evolve to. Depending upon incident variables, the command structure may also vary.

INCIDENT RESPONSE PROCEDURES

The following are provided only as sample methods and procedures that are not intended for use as written. As incidents may vary, response procedures will vary as well. Review the following and before developing organizational emergency response procedures, consider many other resources available from NIOSH, CDC, OSHA, FEMA, NFPA, DOT, and many others.

A valuable resource consideration for all emergency response personnel would be to contact the Department of Health and Human Services, Centers for Disease Control and Prevention (NIOSH) by calling 1-800-35-NIOSH, and request DHHS (NIOSH) Publication No. 2004-103. This is a computer disk that is available at no charge, and it contains many resources of critical hazardous materials emergency response information as contained in at least 10 data bases on the disk.

Sample HAZMAT Emergency Response Guides

GUIDE 1.0 GENERAL HAZMAT INCIDENT RESPONSE:

- Evacuate personnel from immediate hazard area
- Notify site security: about incident, any casualties, request outside support if needed (fire department, medical support & others)
- Contact on-site incident commander
- Contact on-site emergency response team members
- Never respond alone
- Implement on-site incident command system
- Establish perimeter control
- Locate incident command and operations center upwind of incident
- Check vitals of all personnel that will wear PPE
- Assemble first-responder entry team
- First responders enter from the upwind direction
- Assemble first-responder backup team
- Assemble DECON team
- Install DECON Zone, upwind of incident within "warm zone."
- Install first aid and medical support, cold side of DECON corridor
- Assemble and direct Support Teams
- Respond to stabilize the incident, stop source of contamination and handle personnel casualties.
- Decontaminate incident zone
- Clean up incident zone
- Decontaminate entry personnel
- Correct cause of incident
- Determine that incident area is safe for re-entry of workers
- Make notifications to city, state, and federal regulatory agencies as required
- Meet and discuss incident with all responders at close of incident
- Complete, distribute, and file a formal incident report.

Please note: These guidelines must be evaluated and updated to ensure regulatory compliance and to ensure that the most correct and practical incident response guidelines are chosen for use. These guides are provided for informational purposes only. The author makes no warranty, either expressed or implied, with respect to the completeness or continuing accuracy of the information contained herein and disclaims all liability for reliance thereon. Additional information will be necessary for specific conditions and circumstances of use. It is the user's responsibility to determine the suitability of emergency response guidelines and to evaluate risks and precautions required for protecting response personnel and others involved in emergency response operations.

GUIDE 2.0 RESPONSE GUIDELINES

Guide 2.1   Acetic anhydride

Storage: Bulk storage
CAS 108-24-7
Hazard: Acid Corrosive, Fire: Flash Potential LEL = 2.7%
IDLH: 200 ppm
MW: 102.1
PEL: 5 ppm; 20 mg/cubic meter
Flash Point: 120 F
DOT Hazard Class: 8, Corrosive; 3, Flammable Liquid PG II
ERG Info: UN1715, ERG: # 137

Reactive with: water, alcohols, strong oxidizers, amines, and caustics. Forms acetic acid with water; corrosive to human and other living tissues and metals - Highly Volatile

Personal Protection:

- Level B Protection: Hazmat team initial entry, during stabilization, and whenever there is inadequate ventilation; for any large spills (> 30 gallons) and at air concentrations at or above PEL
- Level B with Flash Protection: for incidents involving fire potential; or at or above flash point temperatures (120F)
- Level C Protection: Possible downgrade during final cleanup at air concentrations < PEL, if air contaminant concentrations are within limitations of air purified respiratory protection - use organic/acid air purified respirator cartridges

Additional Considerations:

- Prevent contact with reactive substances
- Eliminate ignition sources
- Have fire suppression and fire fighting equipment nearby
- Do not enter if explosive atmosphere exists or if air concentration exceeds 10% of LEL
- Ensure that PPE is acid resistant
- Pump transfer spilled liquids to acid and grounded fire resistant sealed containers
- Neutralize residual and contaminated materials with soda ash
- Absorb and dispose as hazardous waste if not neutralized or if contaminated with RCRA regulated materials
- Significant releases to the environment require reporting to city, state, and federal regulatory agencies. RQ = 5000 lbs

Guide 2.2   Sodium Hydroxide, Caustic Soda (50%)

Storage: Bulk and 25% Totes
CAS: 1310-73-2
Hazard: Strong Alkali Corrosive
IDLH: 10 mg/cubic meter
MW: 40
PEL: 2 mg/ cubic meter
Flash Point: NA
DOT Hazard Class: 8, Corrosive Alkali, PGII
ERG Info: UN1823 Solid, UN1824 liquid ERG# 154

Reactive with: water, acids, flammable liquids, organic halogens, metals especially aluminum, tin, and zinc. Corrosive to most metals and also reacts with nitro methane.

Low volatility, unless involved in fire or high temperature situations. When reacting with combustible materials it can cause enough heat to ignite the combustible materials.

Personal Protection:

- Level B if involved in high heat or reactive situations
- Level C at ambient temperatures and if air contaminant levels are within limitations of air purified respiratory protection

Additional Considerations:

- Prevent contact with reactive substances
- Ensure PPE is alkali corrosive resistant
- Pump transfer spilled liquids to alkali resistant sealed containers
- Scoop and sweep spilled solids to alkali resistant sealed containers
- Neutralize residual and contaminated materials with vinegar or diluted weak, acetic acid
- Absorb and dispose as hazardous waste if not neutralized or if contaminated with RCRA regulated materials
- Significant releases to the environment require reporting to city, state, and federal regulatory agencies. RQ = 1000 lbs

Guide 2.3   Diesel Fuel, Fuel Oil #2

Storage: Bulk
CAS: 68476-30-2
Hazard: Flammable ignitable liquid
IDLH: (acute LD50 5 ml/kg)
MW: (vapor density = heavier than air)
PEL: 5 mg/ cubic meter
Flash Point: 100 F
DOT Hazard Class: 3, Flammable Liquid PG III
ERG Info: NA 1993 ERG# 128

Reactive with: Open flames or sparks, corrosives, oxidizers, don't use with Viton or Fluorel products. Considered most volatile at or above flashpoint temperatures.

Personal Protection:

- Level B with Flash Protection, large spills (> 30 gallons)
- Level C possible downgrade after free flowing liquids are collected and adequate ventilation has decreased air concentrations below 5% of LEL, and if air contaminant levels are within limitations of air purified respiratory protection

Additional Considerations:

- Prevent contact with reactive substances
- Prevent contact with spilled product
- Eliminate ignition sources
- Have fire suppression and fire fighting equipment near by
- Do not enter if explosive atmosphere exists, or if air concentration exceeds 10% of the LEL
- Pump transfer free liquids to grounded, fire resistant sealed containers
- Absorb and dispose all residual liquids as ignitable hazardous wastes
- Significant releases to the environment require reporting to city, state, and federal regulatory agencies. RQ = Not Listed, however EPA requires reporting of any quantity of oily product that is released to water and produces a "sheen" on top of the water

Guide 2.4   Epichlorohydrin

Storage: 55 gallon drums
CAS: 106-89-8
Hazard: Toxic, Flammable Liquid, carcinogen
IDLH: 75 ppm
MW: 92.5
PEL: 5 ppm, 19 mg, / cubic meter
Flash Point: 93 F
DOT Hazard Classes: 6.1 Toxic, 3, Flammable Liquid PG II
ERG Info: UN 2023, ERG#131

Reactive with: oxidizers, acids, acid salts, caustics, zinc, aluminum, water; can result in hazardous polymerization in presence of strong acids or bases - especially when heated. Consider as a volatile liquid in all environments.

Personal Protection:

- Level A PPE: for large spills and all spills at or above flash point temperatures.
- Flash Protection Required when air concentrations are at or near 5% of LEL 3.8%
- Level B PPE: possible downgrade if air concentrations below PEL and after free liquids are removed

Additional Considerations:

- Prevent contact with reactive substances
- Avoid entering spilled materials
- Eliminate ignition sources
- Have fire suppression and fire fighting equipment near by
- Do not enter if explosive atmosphere exists or if air concentration exceeds 10% of LEL
- Pump free liquids into grounded, sealed ignition resistant containers
- Absorb and dispose all residual and contaminated materials as toxic and ignitable hazardous wastes
- Significant releases to the environment require reporting to city, state, and federal regulatory agencies. RQ = 100 lbs

Guide 2.5   Ferric Chloride Solutions

Storage: Bulk
CAS:
Hazard: Corrosive Acid Chloride
IDLH: ND
MW: 198
PEL: 1 mg/ cubic meter
Flash Point: NA
DOT Hazard Class: 8, Corrosive PG III

Reactive with: strong alkali, oxidizers, corrosive to human flesh and metals. Low volatility at ambient temperatures, volatile and potential HCL and chlorine release at high temperatures.

Personal Protection:

- Level C Protection, having acid resistant protective clothing and if air contaminant levels are within limitations of air purified respiratory protection
- Level B Protection if substance has been involved in fire or reactive substances cause the release of chlorine, or if air contaminant levels exceed Level C protection

Additional Considerations:

- Prevent contact with reactive substances
- Ensure PPE is acid resistant
- Pump transfer spilled liquids to acid resistant containers
- Neutralize residual and contaminated materials with soda ash
- Absorb and dispose as hazardous waste if not neutralized or if containing RCRA regulated materials
- Significant releases to the environment require reporting to city, state, and federal regulatory agencies. RQ = 1000 lbs

Guide 2.6   Gasolines

Storage: Bulk
CAS: 8006-61-9
Hazard: Flammable Liquid, carcinogen
IDLH: ND
MW: 72
PEL: ND
Flash Point: -45 F
DOT Hazard Class: 3, Flammable Liquid PG II
ERG Info: UN 1203 ERG# 128

Reactive with: Strong acids and oxidizers such as peroxides, nitric acid, and per-chlorates.

Personal Protection:

- Level B with Flash Protection for all spills

Additional Considerations:

- Prevent contact with reactive substances
- Eliminate ignition sources
- Have fire suppression and fire fighting equipment nearby
- Do not enter if explosive atmosphere exists, or if air concentrations exceed 10% of LEL
- Stay out of spilled product
- Pump transfer spilled product to grounded, fire resistant metal containers
- Absorb and dispose residuals and contaminated materials as ignitable hazardous wastes
- Significant releases to the environment require reporting to city, state, and federal regulatory agencies. RQ = 3000 lbs, however EPA requires reporting of any quantity of petroleum product that is released to water and produces a "sheen" on top of the water

Guide 2.7   Hydrochloric Acid (33% solution)

Storage: Bulk
CAS: 7647-01-0
Hazard: Acid chloride solution, corrosive to human flesh and metals
IDLH: 50 ppm
MW: 36.5
PEL: 5 ppm and 7 mg / cubic meter
Flash Point: NA
DOT Hazard Class: 8, Corrosive Acid Liquid PG II
ERG Info: UN 1789, ERG#157

Reactive with: hydroxides, amines, alkalis, corrosive to most metals - especially to copper, brass, zinc; corrosive to human tissues; volatile at ambient temperatures.

Personal Protection:

- Level B PPE with acid resistant materials for releases of > 3 gallons
- Level C PPE possible down grade, with acid resistant PPE for small releases in well-ventilated areas and if air concentrations of contaminants are within limitations of air purified respiratory protection. Ensure that air-purified respirators contain acid/organic filters.

Additional Considerations:

- Prevent contact with reactive substances
- Ensure that PPE and materials used are acid resistant
- Pump transfer spilled liquids to acid resistant containers
- Neutralize residuals and contaminated materials with soda ash
- Absorb and dispose non-neutralized material or materials containing RCRA regulated materials as hazardous wastes
- Significant releases to the environment require reporting to city, state, and federal regulatory agencies. RQ = 1000 lbs

Guide 2.8   Hydrogen Peroxide (35% solution)

Storage: Bulk
CAS: 7722-84-1
Hazard: Reactive oxidizer and corrosive liquid
IDLH: 75 ppm
MW: 34
PEL: 1 ppm and 1.4 mg / cubic meter
Flash Point: NA
DOT Hazard Class: 5.1, Reactive Oxidizer and 8, Corrosive Liquid, PGII
ERG Info: UN 2014, ERG: #140

Reactive with: oxidizable materials, combustible materials - may result in spontaneous combustion; also reacts with iron, copper, brass, bronze, chromium, zinc, lead, silver and manganese. Low volatility—prevent formation of reactive peroxide salts & crystals .

Personal Protection:

- Level C PPE with acid corrosion protection and if air contaminant concentrations are within limitations of air purified respiratory protection
- Level B PPE anytime contaminant levels exceed Level C protection

Additional Considerations:

- Prevent contact with reactive substances
- Have fire suppression and fire fighting equipment nearby
- Use corrosion resistant materials and anti-friction equipment (plastics)
- Pump transfer spilled liquids to acid resistant containers
- Absorb residuals with non-combustible absorbents and dispose as reactive hazardous wastes
- Significant releases to the environment require reporting to city, state, and federal regulatory agencies. RQ = 100 lbs

Guide 2.9   Phosphorus oxychloride

Storage: 55 gallon drums
CAS: 10025-87-3
Hazard: Corrosive Acid (breaks down to form hydrochloric and phosphoric acids)
IDLH: ND
MW: 153.3
PEL: .1 ppm and .6 mg / cubic meter
Flash Point: NA
DOT Hazard Class: 8, Corrosive Liquid PGII
ERG Info: UN1810 ERG: #137

Reactive with: water, combustible materials, carbon disulfide, dimethyl formamide, metals. Corrosive to human tissues - consider volatile at all temperatures.

Personal Protection:

- Level B PPE using acid resistant materials for all releases.
- Level C PPE possible downgrade for final clean up, if adequate ventilation to reduce air concentration below PEL and if contaminant air concentrations are within limitations of air purified respiratory protection. Use acid/organic cartridges for air purified respiratory protection.

Additional Considerations:

- Prevent contact with reactive substances
- Use acid resistant PPE
- Pump transfer spilled liquids to acid resistant sealed containers
- Neutralize residual and contaminated materials with soda ash
- Absorb and dispose as hazardous waste, if not neutralized or if waste contains RCRA regulated substances
- Significant releases to the environment require reporting to city, state and federal regulatory agencies RQ = 1000 lbs

Guide 2.10   Propylene Oxide

Storage: Bulk
CAS: 75-56-9
Hazard: Extremely Flammable Liquid, Carcinogen
IDLH: 400 ppm
MW: 58.1
PEL: 100 ppm and 240 mg / cubic meter
Flash Point: -35 F
DOT Hazard Class: 3, Flammable Liquid PG I
ERG Info: UN1280 ERG: #127

Reactive with: anhydrous metal chlorides, iron, strong acids, caustics and peroxides; hazardous polymerization can occur at high temperatures, or when contaminated with alkalis, aqueous acids, amines or acidic alcohols; Irritant and damaging to human tissues.

Personal Protection:

- Level B PPE with flash protection for all releases in areas that air concentrations are at or above 5% of LEL (2.3%)
- Level C PPE downgrade possible at final clean up, having no free liquids and air concentrations are less than 5% of LEL and contaminant levels are below the PEL and within limitations of air purified respiratory protection

Additional Considerations:

- Prevent contact with reactive substances
- Eliminate ignition sources
- Have fire suppression and fire fighting equipment nearby
- Do not enter if explosive atmosphere exists or if air concentration exceeds 10% of LEL
- Pump transfer spilled liquids to grounded, fire resistant sealed containers
- Absorb and dispose all residuals and contaminated materials as ignitable hazardous wastes
- Significant releases to the environment require reporting to city, state, and federal regulatory agencies. RQ = 100 lbs

Guide 2.11   Sodium Hypochlorite

Storage: Bulk and Totes
CAS: 7681-52-9
Hazard: Oxidizing Agent, Corrosive Liquid, Strong skin and respiratory irritant
IDLH: ND (rat LD 50 = 8.91 g/kg
MW: (vapor density = 1)
PEL: ND
Flash Point: NA
DOT Hazard Class: 8, Corrosive, Reactive PG II
ERG Info: UN1791 ERG: #154

Reactive with: heat, strong acids, strong oxidizers, reducing agents, ammonium containing materials, metals, organic solvents; can decompose to release chlorine gas; if involved with fire - fumes may be toxic and irritating.

Personal Protection:

- Level C PPE, using corrosion resistant plastic and rubber clothing and if air contaminant concentrations are within limitations of air purified respiratory protection. Use acid/organic air purified respirator cartridges
- Level B PPE if in contact with reactive agents or anytime contaminant air concentrations exceed limitations of air purified respiratory protection

Additional Considerations:

- Prevent contact with reactive substances
- Ventilate spill area
- Ensure that PPE and response equipment is corrosion resistant
- Dike and contain spilled liquid
- Pump transfer spilled liquids to sealed corrosion resistant containers
- Spill residual can be neutralized using sodium sulfite, bisulfite, or thiosulfite; do not use sulfates or bisulfates
- Absorb residues with clay, sand, or other non-reactive absorbents, dispose contaminated materials as corrosive, reactive hazardous waste
- Significant releases to the environment require reporting to city, state, and federal regulatory agencies. RQ = 100 lbs

Guide 2.12   Sodium Bisulfite (40% Solution)

Storage: Bulk
CAS: 7631-90-5
Hazard: Corrosive, strong irritant and can cause severe burns to exposed human tissues.
IDLH: ND
MW: 104.1
PEL: 5 mg/ cubic meter
Flash Point: NA
DOT Hazard Class: 8, Corrosive PG II
ERG Info: UN2693, ERG: #154

Reactive with: heat, oxidizes to sulfate upon exposure to air, acids, and oxidizing agents. Decomposition products: sulfur dioxide gas

Personal Protection:

- Level C Protection, having adequate ventilation, and using corrosion protective rubber and plastic clothing, and if contaminant air concentrations are within limitations of air-purified respiratory protection. Use acid sulfite, sulfur dioxide protective air-purified respirator cartridges
- Level B Protection anytime that air concentrations exceed limitations beyond air-purified respiratory protection

Additional Considerations:

- Prevent contact with reactive conditions and substances
- Ensure that PPE is corrosion resistant
- Dike and containerize spilled liquids
- Pump transfer spilled liquids to corrosion resistant sealed containers
- Neutralize residual and contaminated materials with soda ash
- Absorb and dispose non-neutralized wastes or if containing any RCRA regulated materials, as corrosive hazardous wastes
- Significant releases to the environment require reporting to city, state, and federal regulatory agencies. RQ = 5000 lbs

Guide 2.13   Sulfuric Acid (93%)

Storage: Bulk
CAS: 7664-93-9
Hazard: Strong Acid Corrosive and Toxic, serious burns and destroys human tissues
IDLH: 15 mg / cubic meter
MW: 98.1
PEL: 1 mg / cubic meter
Flash Point: NA
DOT Hazard Class: 8, Corrosive, 6.1 Toxic, PG I
ERG Info: UN1831, ERG: #137

Reactive with: water, organic materials, chlorates, carbides, fulminates, powdered metals, corrosive to most metals, and generates enough heat to ignite combustible materials.

Personal Protection:

- Level A Protection: Total encapsulation with corrosion protective plastics and rubber materials for responding to reactive spills that generate fumes, vapors, mists and other air contaminants.
- Level B with Flash Protection: Total coverage protection with corrosive protective plastics and rubber materials and covered with an outer flash protective suit. For responding to reactive spills that have potentials to cause fire.

Additional Considerations:

- Prevent contact with reactive substances
- Do not enter spilled liquid and avoid contact
- Have fire suppression and appropriate fire fighting equipment near by
- Pump transfer spilled liquids to acid resistant sealed containers
- Neutralize residual and contaminated materials with soda ash
- Absorb non-neutralized acids and materials with non-combustible, non-reactive absorbents and seal in acid resistant containers for disposal as corrosive and toxic hazardous wastes
- Significant releases require reporting to city, state, and federal regulatory agencies. RQ = 1000 lbs

APPENDIX

The following are examples of elements contained in a typical Industrial Emergency Program.

SAMPLE EMERGENCY PLANS

Purpose

The purpose of this plan is to describe actions to be taken in the event of an emergency that occurs at the Any Company, general offices located in XXXXXX. Personnel and property covered by this plan are located within the confines of every entity considered Any Company properties and Any Company job sites.

Definition of Emergency

Any condition or incident that may have an impact upon employee safety, health, or significant loss of property and requires immediate action to correct, is considered an emergency. Emergencies may include:

- Personnel injury, illness, and/or death
- Fire, explosion
- Structural collapse
- Natural disaster including tornado, earthquake, floods
- Hazardous substance spills or releases
- Bomb threats and or acts of terrorism
- Any action requiring personnel evacuation
- Any action requiring implementation of Site Contingency or Emergency Action Plans

General Procedure

In the event of facility or job site emergencies, all employees that are not assigned to perform emergency actions will evacuate to safe locations as directed in this plan. Those persons that have been assigned emergency responsibilities and have been properly trained and have required resources will respond as directed in this plan.

Responsibilities Under This Plan

Personnel assigned to perform designated emergency functions are responsible to respond to emergencies as designated below:

- FACILITY MANAGER/CHIEF OFFICER—This person is responsible to ensure implementation of actions described in this plan and to ensure that updates and changes are made to this plan when pertinent information is provided from competent resources. The Chief Officer is responsible for assuring that emergency action activities occur in parallel with hazardous waste contingency, regulated spill contingency and emergency response plans that are relative to the Any Company. The Chief Officer is responsible to appoint an Emergency Coordinator for the Headquarters Facility and to empower an individual to coordinate with Job Site Managers to ensure that all emergency actions are properly handled. This person is responsible to ensure that Site Security is coordinated with internal on-site and job site emergency action resources. Designated personnel listed in this plan are to act in place of the Site Emergency Coordinator in the event that the Site Emergency Coordinator cannot be contacted. The Facility Manager is responsible to report incidents and related issues to Corporate Management. In the absence of the Facility Manager the Assistant Chief Officer will assume the above mentioned responsibilities.

- ASSISTANT CHIEF OFFICER—In the absence of the Chief Officer this person will direct actions described in this emergency action plan and the company hazardous waste management plan. The person that functions in this capacity will implement actions described in this Any Company emergency action plan, spill contingency plans, and emergency response plans respectively. Responsibilities also include notifying the U.S. Occupational Safety and Health Administration and families of employees that are injured or killed on site. Note: In the event that two (2) or more employees are injured and sent off-site for emergency medical treatment, or in the event that one (1) or more employees are killed as the result of work-related activities—the Region X Office of the U.S. Occupational Safety and Health Administration must be notified immediately.

- COMMUNICATIONS COORDINATOR—This person is responsible to maintain up-to-date listings of home and other emergency telephone numbers for company employees. The Site Communications Office shall have copies of this list. Under normal conditions the Communications Office is located at the Any Company Headquarters located at XXXXXX. In any circumstance that may require relocation of the Communications Office, the Chief Officer and Assistant Chief Officer shall be immediately notified of the new location. Communications performed on site shall include use of telephones, portable radios, and other intercommunication equipment. The Communications Officer will coordinate use of communications equipment as conditions may warrant. The Communications Officer shall coordinate incoming and outgoing communications in the event of emergency actions, taking note to ensure that priorities are assigned as necessary. The Communications Officer may perform the following duties during an emergency:

- Coordinate with the company Emergency Coordinator
- Must be knowledgeable and able to work within the company incident command system whenever it is implemented
- Dispatch personnel to incident areas
- Coordinate between incident response personnel and site management
- Monitor and coordinate messages sent out and received on-site
- Make contact with off-site emergency services such as medical, fire department, hazardous substance spill response agencies, local emergency planning committee participants, and regulatory agencies
- Under the direction of the Chief Officer, shall communicate information with families of employees, the community, and to the media

- EMERGENCY COORDINATOR—This person is responsible to coordinate emergency action activities and emergency response actions that occur at Any Company Headquarters and Job Site Locations.
  These activities will include:
  - Coordinate implementation of the company Incident Command System whenever emergency conditions occur
  - Coordinate with the company Chief Officer, Assistant Chief Officer and Communication Officer and Project Managers to ensure that implementation and maintenance of the written Emergency Action Plan occurs, that it is functional and that deficiencies when found in the program are corrected
  - Coordinate with out-of-company mutual aid agencies such as public emergency services, fire services, emergency medical services, regulatory agencies, regulated substance cleanup teams, contractors and others to ensure that Any Company personnel and interests are properly supported when needed
  - Coordinate emergency reporting actions and ensure that proper notifications are given to appropriate regulatory agencies when necessary
  - Monitor routine company operations to ensure that functional emergency action programs are in place and are appropriate for each job site
  - Monitor training activities to ensure that Any Company personnel are appropriately trained to safely perform any required responsibility. This may include emergency action, incident response, cleanup, decontamination, remediation, and other responsibilities

- FIRST AID COORDINATOR—This person is responsible to coordinate on-site first aid activities. This person shall work with the Communications Officer and Emergency Coordinator to interface with off-site emergency medical support agencies. Examples of first aid activities include:
  - Coordinate first aid training for all applicable company personnel

- Develop, implement, and maintain the company first aid program
- Make provisions to safely conduct first aid services within all areas that company personnel may be involved, including situations that require implementation of the company incident command system
- Interface with Any Company emergency response personnel and community emergency response agencies to properly handle and decontaminate injured personnel
- Manage and coordinate services performed by company first aid teams

- SITE SAFETY AND HEALTH OFFICER—This person is responsible to ensure that work activities and emergency activities are performed safely at Any Company properties and job sites, and that they are performed in compliance with all environmental, health and safety regulatory directives. Duties include:
  - Monitor job sites to ensure that proper standard operating procedures are in place and are being followed by company and contractor employees
  - Ensure that proper personal protective equipment is available and is being used by all applicable employees
  - To coordinate with the company Emergency Coordinator to ensure company site safety and health programs are followed by Any Company personnel and by authorized contractors
  - To ensure that all elements of the company safety and health and program are complied with under normal working conditions as well as during emergency conditions
  - To ensure that the Any Company meets regulatory requirements under normal working conditions as well as during emergency conditions

- MAINTENANCE COORDINATOR—This person is responsible to coordinate routine and emergency maintenance activities performed by on-site personnel. This person shall work with the Emergency Coordinator to monitor and to coordinate maintenance activities performed by contractors and other off-site personnel that may respond to emergency conditions on Any Company properties and job sites. Emergency maintenance activities include:
  - Provide support to repair and maintain emergency equipment
  - Provide equipment maintenance support during emergency operations
  - Coordinate with all emergency operations officers as conditions warrant
  - Provide equipment maintenance support during post emergency and production restart operations
  - Provide support to restore communications, utilities and roadways within the facility after incidents occur

- UTILITIES COORDINATOR—The Utility Officer is responsible to work with the Maintenance Coordinator and Emergency Coordinator to perform emergency activities relating to on-site utilities and to support emergency activities that may require utility adjustments at company work sites. Emergency utilities activities include:
  - Controlling energy sources to utilities in support of emergency operations
  - Coordinating with emergency officers to control water resources
  - Making provisions for using auxiliary equipment

Evacuation

Any Company employees are not expected to respond to fight structural fires or to respond to hazardous materials emergencies. In the event of fire or other incidents that may pose a threat to employee safety, all non-emergency response employees are expected to evacuate to safe locations by following evacuation routes posted at each respective job site. After personnel evacuation to a safe location, they are expected to make emergency calls by dialing 911 and also calling company emergency personnel.

Supervisors

Supervisory personnel are responsible to ensure that all employees, contractors and visitors located within their areas of responsibility, are trained to follow applicable emergency procedures listed within this plan. Supervisors, subordinates, and visitors are required to support designated emergency action personnel and to follow procedures as described in this plan.

Any Company Employees and Contractors

All employees and contractors that are not designated to perform emergency action functions are expected to do the following in the event of an emergency:

- Evacuate from incident areas to safe locations
- Alert other employees of potential danger and tell them to evacuate
- At a safe location call the XXXX Headquarters Office and report the incident and provide the following information:
  1. Incident location
  2. Personnel injuries (known & potential)
  3. Nature of the incident: kinds of injuries, fire, chemical spill, odors, known dangers
  4. Materials involved in the incident

1A. EMERGENCY PROCEDURES

1A.1 General Evacuation Procedure

In the event of an incident that warrants evacuation of employees and other inhabitants from Any Company properties and job sites, alarms for evacuation and emergency action instructions shall be communicated. This shall occur by verbal evacuation orders given by persons having knowledge of emergency conditions, by broadcasts, and by other forms of alarms. All persons that do not have emergency action responsibilities shall evacuate from emergency conditions:

- Immediately move to designated emergency exits. Building exits are designated for each floor zone in this plan. Floor plans are posted within each zone on each floor in each building - area supervisors are responsible to ensure that all employees and visitors are familiar with their respective primary and secondary evacuation routes.
- Job site evacuations will vary and safe exit should be discussed with workers prior to beginning work at each respective site.
- Primary emergency exit routes shall lead to designated assembly locations. In the event that primary exits are involved in the incident and cannot be safely followed, move to secondary exits and immediately leave the building. Supervisors are responsible to designate primary and secondary exits and employee assembly locations.
- Handicapped or injured persons must be assisted by at least 2 employees and taken to the nearest, safe, non-congested evacuation point. Supervisors must assign employees to assist persons needing evacuation help. In the event that primary evacuation points cannot be safely accessed, alternate safe evacuation methods shall be used.
- In buildings, floor watchers should be assigned and are responsible to assist employees by directing them to designated evacuation routes. When all employees have been evacuated, or if any remain, floor watchers are to move to a safe location and call the Emergency Coordinator, or Project Manager if Emergency Coordinator cannot be reached, and report zone clear or not clear - then safely evacuate to assembly locations.
- All Supervisors or Project Managers shall notify the Emergency Coordinator and Communication Officer and Emergency Response Incident Commander if any employee is not accounted for or cannot be evacuated for any reason.
- Shipping Dock Attendants, if docks are not involved in an emergency and if it is safe to remain in the area, open emergency exit doors and clear the area to facilitate entry by the Fire Department or other emergency services. Prevent unauthorized entry by anyone, except emergency response personnel, through the dock. Assist in the evacuation of injured or handicapped personnel if properly trained to do so.

1A.2 Emergency Evacuation Drills

Emergency evacuation drills are conducted periodically under the direction of the Facility Manager (Chief Officer). Prior to evacuation drills, the Emergency Coordinator updates all persons having emergency action responsibilities with information related to plan drills.

In-Company Emergency Response Team

If the Any Company chooses to support an in-company emergency response team program, the program shall follow applicable regulatory directives, reference 29 CFR 1910.120 (q) and 29 CFR 1926.65 (q). The Spill Response Team Senior Response Official shall ensure that team members have proper training, supplies, and equipment and coordinate at least 8 hours per year of refresher training for Response Team Members. The company Senior Response Official shall update team members with information relative to emergency action procedures and evacuation drill information as well.

1B EMERGENCY PROCEDURES

1B.1 General Fire Response Procedure

Employees are not expected to fight fires. Persons that choose to fight fires on Any Company Properties and Job Sites shall be properly trained, fight only non-structural fires, have proper firefighting equipment immediately on hand and be able to respond safely. In the event of a fire, or potential fire incident, the following shall occur:

1. All persons in the immediate area must evacuate to safe locations listed on evacuation floor plans.
2. A witness shall notify Services by calling 911 and also notify the company Emergency Coordinator and provide the following information:
   - Location of the fire (floor, column, building)
   - Number of injuries
   - Materials involved
   - Other significant information
3. Evacuate and meet at assembly locations
4. Emergency Coordinator:
   - Sounds alarm and notifies the Emergency Coordinator
   - Coordinates with off-site fire department if required
   - Initiates incident command system, establishes perimeters and site security actions
   - Coordinates with Spill Team Senior Response Official to initiate any required support actions

- Coordinates with Maintenance and Utilities Coordinators to remove smoke and fumes from buildings after emergency conditions are stabilized

5. Emergency response team and/or off-site Fire Department respond to stabilize the fire and other emergency conditions related to the incident.

6. The Emergency Response Team Senior Response Official, and Emergency Coordinator interface with the off-site Fire Department Incident Commander to evaluate personnel casualties, to implement recovery operations, to implement decontamination actions and to prepare for safe re-entry into incident areas.

7. In-company Emergency Response Team members are to assist the local Fire Department and other emergency response agencies in performing air quality checks, decontamination, and to prepare incident areas for safe re-entry of support and Any Company personnel and contractors.

1B.2 General Chemical Spill Response Procedure

In the event of a spill or release of hazardous or regulated substances, the following shall occur:

1. All personnel in the immediate area must evacuate to safe locations.
2. A witness to the incident must notify the Emergency Coordinator and provide the following information:
   - Location of incident (floor, column#, building)
   - Number of injuries
   - Materials involved
   - Potential for fire or explosion
   - Other significant information
3. The Emergency Coordinator and Spill Team SRO implement perimeter control and incident command procedures.
4. Spill Team members or local fire department team members are notified by the Emergency Coordinator or Senior Response Official and are instructed to assemble at a safe location to prepare for responding to the incident.
5. Team responds to the incident area to stabilize the incident, decontaminate the spill area, make needed notification, and to prepare the incident area for safe re-entry of personnel.

1B.3 General Response to an Earthquake

In the event of an earthquake, the following must occur:

- Be calm and remain in the immediate area.
- Get out of elevators and stairwells.
- Move from exterior walls, skylights, overhanging fixtures or objects that may fall.

- Take cover under a sturdy structure (desk, table, door frame).
- Do not evacuate the building until directed to do so by the fire marshal.
- When evacuation signal is sounded, follow general evacuation procedures as previously listed in this plan.
- Plan for after shocks during post evacuation assembly. Assembly points must be located in safe locations as listed in item 3.
- Use caution and avoid electrical hazards, hazardous substance spills and releases and other hazards.

CAUTION—If riding in a motor vehicle do the following:

- Pull over, stop the vehicle, and remain in the vehicle.
- Do not park under bridges, overpasses, or electrical wires
- Remain in the vehicle and avoid all fallen electrical wires.
- Avoid crossing damaged bridges or overpasses.
- Avoid fallen electrical wires. Remain in vehicle.

1B.4 General Response to Tornado

In the event of a tornado warning the following must occur:

1. Immediately move to inward corridors of buildings away from windows, doors, and outside walls.
2. Dock attendants instruct drivers to move trucks away from dock areas and close doors.
3. Move to assembly points located within designated tornado shelter areas:
   - Buildings
   - Floors
   - Basements

Note: Emergency Response Teams and Security members are to assemble at the Security/Communication Center Checkpoint if safely possible. In the event of damage to Any Company Buildings from natural disaster, fire, explosion or hazardous substance release, Site Security, company Emergency Response Team members and Community Emergency Response Agencies will assemble to perform recovery operations and to evaluate damages prior to allowing re-entry of personnel.

1B.5 Emergency Notification Procedures

Death or Multiple Injuries
In the event of one (1) or more personnel death, or injury of (2) or more personnel, and the injuries require medical treatment, the Regional 7, Occupational Safety &

Health Administration (OSHA) must be notified immediately. Any death shall be reported to the local emergency services, including police department, by calling 911. This should be done as follows:

- In the event of any emergency, the Emergency Coordinator must be notified and personnel injuries (and death) must be reported to the Facility Manager or Chief Officer. The Facility Manager or the Emergency Coordinator shall immediately notify local emergency response agencies, OSHA, company representatives, and other agencies as required.
- Site Security or the Emergency Coordinator also notifies Emergency Medical Services on-site and off-site when required. (Call 911 when off-site medical support is needed.)

HAZARDOUS SUBSTANCE SPILLS OR RELEASES

Immediate notification is required in the event of hazardous substance spills or releases that leave the Any Company Properties or Job Sites by way of the air, sewer, and water drainage or into the land. Notification of spills and releases must be made to Region X EPA, state and local air quality or local city waste water treatment, depending upon the mechanism of release. Immediate notification of spill or release must be made to Site Emergency Coordinator and Facility Manager by the following procedure:

- All emergencies, including hazardous substance spills or releases must be reported to the Emergency Coordinator.
- The Emergency Coordinator notifies the company Spill Team Leader or community emergency services and actions occur to handle the incident.
- The Spill Team Leader is required to coordinate activities and reportable issues with the Emergency Coordinator.
- The Emergency Coordinator is required to immediately notify off-site support agencies and regulatory agencies as required, to receive support, or to report hazardous substance releases. The Emergency Coordinator shall further investigate and correct the cause of the incident after the emergency is stabilized.

THREATS OF TERRORISM

Appropriate action should be taken in each incident involving a bomb threat or suspected threat of terrorism. When a caller indicates a bomb threat, or an act of terrorism is suspected, the person answering the call should:

- Tell the caller, "We'll have a manager get with you on this."

- Ask for name and address of caller.
- Ask questions and take note of information.
- Site Security, the Emergency Coordinator, and the Facility Manager must be immediately notified and further action shall be taken—conduct employee evacuation, initiate law enforcement action, search etc.
- Report all bomb threats and other terrorist threats to appropriate local law enforcement agencies and to the FBI.

TRAINING & PARTICIPATION

Any Company Supervisors are responsible to ensure that all employees under their direction are properly trained to follow emergency action procedures as they may apply. Supervisors should encourage employee participation in emergency action and safety programs.

This should be done as follows:

- Provide training in safety meetings
- Inform new employees before beginning work
- Enforce compliance and participation in safety drills
- Coordinate with the Site Emergency Coordinator, Floor Watchers, and other Emergency Action Officers
- Discuss drill results and incidents with employees
- Provide initial, on-going, and annual Hazard Communication and Emergency Action Refresher Training for all applicable employees
- Provide Initial HAZWOPER and Annual HAZWOPER Refresher Training for all applicable employees having emergency action responsibilities
- Encourage department representatives to participate on safety committees, emergency evacuation teams, and emergency response teams

Chapter 6

CORROSIVE SUBSTANCES

Figure 6.1 What are corrosives?

WHAT ARE CORROSIVE HAZARDS?

Using common sense terms, corrosive hazards are liquids, vapors, gases, or solid materials that cause visible destruction to human tissues or are substances that dissolve metal. Basically if a substance eats up your skin or dissolves your car, it is very likely corrosive.

Because most chemical spills and releases occur during handling of containers and during transportation, the United States Department of Transportation (D.O.T.) regulates corrosive materials using the following criteria:

1. A corrosive material is considered destructive or can cause irreversible alteration in human skin. In this definition, the structure of tissue at the site of contact is destroyed or changed irreversibly after an exposure period of four hours or less. Note: Who would wait four hours? Unfortunately, sometimes this happens. Prevent exposure, but if you become exposed? Immediately wash skin and remove clothing that has been exposed to corrosive substances.

2. A substance is considered to have severe corrosion properties if it dissolves 0.250 inches per year on steel at 130 degrees F. This is based upon the Society of Safety Engineers criteria listed in the standard SAE 1020.

As mentioned in Chapter 3, the U.S Environmental Protection Agency (EPA) regulates a corrosive waste if it has an acid pH of pH2 or less and an alkali pH of 12.5 or higher. The D.O.T. characterization of corrosive materials more stringently suggests that materials having properties and pH values at either side of EPA hazardous waste criteria are significant.

About Corrosive Liquids

From a safety standpoint, corrosive liquids are most commonly involved in incidents that cause personnel injury, spills, and releases to the environment. Corrosive liquids most often expose the skin and eyes. The mineral acids, acid organics, caustic bases, and some organic solvents are considered corrosive liquids. Hazardous vapors are often emitted from corrosive liquids such as vapors or mists from ammonia, nitric acid, hydrochloric acid, acetic acid, bromine and several other substances.

About Corrosive Solids

Dusts, crystals, and powders from corrosive substances are less acute in potential harm than corrosive liquids, vapors, and gases. Workers that deal with corrosive solids must ensure that they use respiratory protection. Humidity and moisture on the skin play a key part in dangerous potentials that exist from being exposed by corrosive solids. When corrosive solids are exposed to moisture on the skin, exothermic reactions occur, heat is released, and further damage occurs to the tissues. Corrosive products such as sodium hydroxide, potassium hydroxide and certain sulfides are common contributors to injuries that occur from corrosive solids.

About Corrosive Gases

Corrosive gas products should be considered extremely hazardous. Gaseous corrosive substances can be quickly absorbed into the body by absorption and by inhalation. Some examples of corrosive gases would include anhydrous ammonia, hydrogen chloride, hydrogen fluoride, and formaldehyde.

What about regulation of corrosives by the U.S. Environmental Protection Agency and State Environmental Agencies?

As mentioned earlier, the Environmental Protection Agency (EPA) uses pH as the criterion for determining if a substance is corrosive, again from waste management perspectives.

The EPA corrosive waste definitions suggest using the following criteria:

- Aqueous or liquid like substances having acid pH values of 2 or less.
- Aqueous or liquid like substances having alkali pH values of 12.5 or higher.
- Also, a waste material that will corrode steel at ¼ inch per year.

Concentration of the corrosive product contained in a substance is a determinant of its strength and ability to sustain damage. From a waste disposal perspective, the corrosive concentration has an impact upon methods used to treat and dispose the waste. Generation and disposal of corrosive wastes is tracked on documents required by the Department of Transportation, Environmental Protection Agency and state environmental regulatory agencies. The EPA tracking code for wastes that are categorized as corrosive is D002.

Use of Corrosive Materials

Thousands of corrosive materials are used daily throughout the world. Some materials can be determined as corrosive easily by name and others are not so obvious, review the products listed in Figure 6.3.

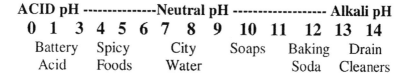

ACID pH ---------------**Neutral pH** ------------------ **Alkali pH**

0	1	3	4	5	6	7	8	9	10	11	12	13	14
Battery Acid			Spicy Foods			City Water			Soaps		Baking Soda	Drain Cleaners	

Figure 6.2   Example of pH ranges from 0 to 14.

COMMONLY USED	CORROSIVE SUBSTANCES
Acetic Acid	Ammonium Hydroxide
Chromic Acid	Hydrobromic Acid
Hydrochloric Acid	Hydrofluoric Acid
Nitric Acid	Perchloric Acid
Phosphoric Acid	Potassium Hydroxide
Sodium Hydroxide	Sulfuric Acid
Acetic Anhydride	Chlorosulphonic Acid
Hydriodic Acid	Phosphoric Pentoxide
Proprionic Anhydride	Aluminum Chloride
Ammonium Difluoride	Antimony Trichloride
Antimony Pentachloride	Bromine
Phosphorus Oxychloride	Phosphoryl Chloride
Phosphorus Trichloride	Phosphorus Pentachloride
Potassium Difluoride	Sodium Difluoride
Sodium Hypochlorite	Stannic Chloride
Sulfur Tetrachloride	Sulfuryl Chloride
Thionyl Chloride	Titanium Tetrachloride
Zinc Chloride	Acetyl Chloride
Allyl Iodide	Benzoyl Chloride
Benzylamine	Benzyl Chloroformate
Chloroacetyl Chloride	Ammonium Polysulfide
2-Chlorobenzaldehyde	Hydrazine
Hydrogen Peroxide	Silver Nitrate

Figure 6.3 Commonly shipped corrosive substances.

PROTECTION FROM CORROSIVE HAZARDS

In Chapter 3, we discussed properties of hazardous materials. In terms of corrosive hazards, we should consider such properties of volatility and vapor density. By knowing that a corrosive product will rapidly vaporize, it is important that we know where the vapors or gases will go. Depending upon air flow, heavier than air, corrosive emissions are likely to remain within lower portions of room air space; likewise, lighter vapors will be found in elevated portions of a room.

How should corrosive emissions be controlled? Corrosive mists, vapors, and gases should be ventilated and hopefully treated before they enter the atmosphere. How should ventilated corrosive emission be treated? By filtering, washing and neutralization before the emissions leave the ventilation exhaust system.

Ventilation

The Occupational Safety and Health Administration provide guidance for protecting employees from corrosive and other hazardous emissions. Protection from corrosive emissions should also focus upon personal protection and nearby emergency equipment such as emergency showers. A serious exposure concern is for people that may be around open spills or may work near open tanks. For guidance in this regard, refer to OSHA Standards 29 CFR 1910.94; 29 CFR 1910.111; 29 CFR 1910.261; 29 CFR 1910.262 and 29 CFR 1910.268.

Personal Protection

As previously mentioned, the best way to keep from being injured by corrosive materials would be to keep them from contacting your body. Remember, we need to protect ourselves by preventing dangerous materials from entering the 4 modes of exposure outlined in Figure 6.4.

Choices made for using personal protective equipment must be made after knowing what hazardous properties a corrosive material may have. Isn't knowing that a substance is corrosive enough information? No.

Some corrosive materials may penetrate "one type works for all" kinds of PPE. Use reference information when choosing the proper personal protection. Material safety data sheets, NIOSH Pocket Guide for Chemical Hazards, PPE Manufacturers Guides, internet resources and many other references are available to keep us from making a mistake by making hasty assumptions.

For more information about proper protection against corrosive hazards, refer to OSHA Guidance in 29 CFR 1910.132, 29 CFR 1926.28, 29 CFR 1910.133 and 29 CFR 1926.102

Typical chemical protective clothing used when handling corrosives:

- Rubber gloves
- Aprons
- Face Shields
- Plastic and Rubber Personal Protective Clothing
- Respirators
- Protective shoe coverings

4 Modes of Exposure

1. Inhalation, which is the fastest and most common means of exposure.
2. Absorption through the skin and tissues.
3. Penetration of skin, entry into the blood stream by puncture, cuts, and openings in the skin.
4. Eating, or Ingesting, dangerous substances.

Figure 6.4 Modes of exposure to dangerous substances.

- Level A total encapsulation personal protective equipment
- Level B total coverage personal protective equipment
- Level C total coverage personal protective equipment

Corrosive Hazard Protective Measure Checklist

The following should be considered for corrosive substance work areas.

- An official documented personal protective equipment evaluation on file.
- Ventilation and engineering controls to reduce PPE requirements.
- Specific protection of skin, eyes, and respiratory system.
- Overall body protection including, arms, hands, feet, head and face.
- Eyewash and safety showers
- Personnel trained to choose and don personal protective equipment.
- Personnel trained to doff and decontaminate personal protective equipment.
- Personnel trained in corrosive exposure first aid procedures.
- Personal hygiene, wash up facilities located near PPE use area.
- Corrosive systems maintained in controlled environment.
- Incompatible substances not used or stored near corrosive substances.
- OSHA Hazard Communication labeling and marking of process and containers.
- Neutralization kit near by for small spills or leakage of corrosives.

Figure 6.5 Toxic properties of corrosive substances.

- Emergency response program in place for corrosive spills or emergencies.
- Fire prevention provisions in place.
- Corrosion resistant process equipment.
- Corrosion resistant facility and utilities.
- Corrosion resistant storage containers.
- Secondary containment for process and storage areas.
- All materials stored off the floor to allow clean up and neutralization of spillage.
- Closed loop and safe corrosive substance transfer systems.

Figure 6.6 Responders overhaul a damaged pressurized cylinder.
(Courtesy of Joesph Hilliard and HMTRI/CCCHST)

Corrosive Air Contaminants

All the properties that we discussed in Chapter 3 and especially those considerations that relate to air emissions containing corrosive materials need to be revisited. Several years ago in California, an automobile dealer noticed that any car that remained on his lot longer than one week, needed a new paint job. Beautiful chrome finishes on the new cars rapidly became pitted. It appeared that after a precious metal recycler moved into the neighborhood, these problems occurred. Oh yes, the neighbors all had runny noses and red eyes as well. The recycler was using a process to dissolve gold, platinum and rhodium that contained nitric acid, hydrochloric acid and sulfuric acid. The red brown fumes spread through the surrounding environment, as the recycler operated in the darkness of night. It didn't take long for the authorities to pinpoint the recycler, thousands of dollars and several injured people later.

Combinations of corrosive products will produce increased hazardous properties that may not be considered when choosing standard rubber or plastic personal protection. Also corrosive fumes, vapors, mists, gases, and dusts may present special conditions and a requirement for specific, highly resistant respiratory protection equipment. Know the hazardous properties of materials that you may be working with. For more information on respiratory protection from corrosive hazards, refer to OSHA 29 CFR 1910.134, 29 CFR 1926.103 and 29 CFR 1910.262.

Health Effects from Corrosive Substances

The most common hazardous exposures produced by corrosive products are to the skin, eyes, digestive, and respiratory systems. Exposures to corrosive substances primarily occur from poor ventilation near chemical processes, and performing improper procedures when opening, pouring and similar types of handling.

Most severe exposures occur when very concentrated corrosive liquids, vapors, or gases have an opportunity to affect the body by way of the 4 modes of exposure, see Figure 6.4. Of course, the longer a corrosive substance remains on body tissues the most severe damage occurs.

Alkali chemicals cause human tissue to form a clot, sometimes called an albuminate. This almost resembles soap like substance. An example of damage that could occur to the eye would be demonstrated by dripping a few drops of caustic soda onto a raw egg. Human tissue will become like gelatin and corrosive damage will seep painfully into the tissues. Acids have a tendency to harden the skin and produce localized pain at the site of contact. In this regard, acids may produce faster pain response than alkali chemicals.

What should an exposure victim do? As routinely reported in most MSDS, flood the point of exposure with lots of clean water for at least 15 minutes. Respiratory exposures are extremely serious. Corrosive vapors, mists, or gases will damage or irritate the lungs and respiratory system in such a manner that an often fatal pneumonia like reaction occurs. All portions of the human and other animal bodies are subject to severe damage from corrosive substance exposure.

Monitoring Potentials for Corrosive Exposure

Volatile corrosive chemical processes require ventilation and routine monitoring. Of course, all personnel involved in this type processing should wear eye protection as well as protection for any portion of the body that may become exposed during process operation.

Where are corrosives found within industry?

Acids & Alkalis

- Metal Etching
- Electrolytic Cleaning
- Metal surface finishing and pickling
- Electroforming and Electroplating
- Bulk Food Preparation
- Water Treatment
- Waste Treatment
- Battery recycling
- Metal recycling
- Chemical manufacturing
- Paper making & recycling
- General Manufacturing & Building Maintenance

SAFE HANDLING OF CORROSIVES

Corrosive materials are found as liquid, solid and gas. There are some substances that become corrosive when they become in contact with moisture. This property is a strong concern for personnel that may handle corrosive salts and other materials. Again, the use of corrosion resistant chemical protective clothing is warranted in this situation.

As previously mentioned, several corrosive substances are used in industry and agriculture that may considered corrosive or may become corrosive to human tissue and metals.

They include:

- Acids and anhydrides
- Alkalis (bases)
- Halogens (chlorine, fluorine, bromine, iodine) and halogen salts
- Organic halides, organic acid halides, esters and salts
- Miscellaneous corrosive substances (substances that are widely used but do not fall into the above mentioned classes)

Over 22 percent of chemicals used in the world are produced in the United States. Some of the highest quantities of chemicals produced are corrosive, such as:

- Sulfuric Acid
- Ammonia
- Phosphoric Acid
- Sodium Hydroxide (caustic soda)
- Nitric Acid
- Hydrochloric Acid

It must be mentioned that the highest number of dangerous chemicals that are spilled or released in the United States along with petroleum products, are also corrosive. The corrosives most commonly spilled or released are: ammonia, sulfuric acid, chlorine, and hydrochloric acid. Hundreds of personnel are trained each year in the United States to respond to chemical spills, especially in response to dangerously volatile corrosives substances.

An extremely important training center that provides training to many hazardous materials training professionals in the United States is the Hazardous Materials Training and Research Institute, in Cedar Rapids, Iowa (see figure 6.7).

Hundreds of hazardous materials training professionals are trained in the United States each year.

Common Sense Procedures for Handling Corrosives

Handlers of corrosive products should wear corrosion resistant aprons, gloves, and splash protection (includes face shields and tight fitting chemical resistant goggles).

- Corrosives contained in bottles should be transported in beak resistant bottle carriers, especially for containers that are larger than 500 ml.
- Never transport corrosive containers by holding on to plastic bottle rings - used for controlling pouring of corrosive liquids.
- Only a fool would pipette corrosive liquids by mouth - don't be a fool... Always use a pipette bulb or other kinds of pipette devices.
- Avoid breathing corrosive vapors and never try to smell corrosive substances
- Always tightly close containers of corrosive substances.
- When mixing or blending corrosive substances, always slowly pour the substances together.
- When mixing corrosive substances with water, always add the corrosive substance to the water - never add water to the corrosive substance.
- Beware of containers used for blending corrosive substances.

Figure 6.7 Hazardous materials trainees conducting an emergency response drill for dealing with corrosive substances. *(Courtesy of Joesph Hilliard and HMTRI/CCCHST)*

- Follow the OSHA Hazard Communication Standard - always clearly label containers that contain corrosive substances using OSHA guidelines such as listing the name, hazard, protection and handling requirements, target organs affected by the product and similar information.

Corrosive Substance Storage

- Do not store corrosive substances near materials that will react with them.
- Never store flammable or toxic substances near corrosives.
- Do not store acids and bases together.
- Separate corrosive liquids from corrosive solids.
- Keep corrosives in corrosion resistant storage cabinets.
- Corrosive products should be stored on plastic trays or shelves.
- Never store corrosive substances over head.
- Do not store organic acids with strong oxidizing agents: What are they?
  1. Examples of strong organic acids include: formic acid, acetic acid, acetic anhydride, and the like.
  2. Examples of strong oxidizing agents include: sulfuric or nitric acid.

Note: When reactive substances such as strong organic acids and strong oxidizers are stored together, dangerous vapors are chronically emitted. Not only will they present hazards for personnel, they also eat up all metal fixtures within the storage room.

Bulk Storage of Corrosive Substances

Tanks used for storing corrosive substances require the most stringent of construction requirements. In this regard, the substance density and other dangerous physical properties affect tank construction components and standards. Of course materials used to construct a corrosive substance storage tank must be corrosion resistant. All related hardware and fixtures must also be corrosion resistant.

Steel and stainless steel is commonly used for making corrosion resistant tanks and storage units; however, some types of steel will not resistant corrosive properties. Carbon steel and other kinds of steel will not resist many corrosive substances. In this regard, some tanks may be coated with ceramic, glass, plastic, resistive metals and paint.

Aluminum tanks are limited in the types of corrosive substances that they may hold. In this regard, most aluminum tanks are not used for holding acid or alkali chemicals.

Heavy duty, reinforced plastics, and resin wound fiber glass containers are often used for making bulk storage containers for corrosive substances. Again, there may be certain types of corrosive materials that may weaken polymer and plastic tanks. Materials assessments must be made before investing in equipment used for holding corrosive substances. The most common plastic and polymer tanks are constructed using:

- Polyvinyl chloride (PVC) plastic
- Polyethylene
- Polypropylene
- Polytetrafluorine ethylene

All corrosive storage tanks require marking as required by the OSHA Hazard Communication Standard. They require testing, routine maintenance, and inspection. Relative documents should be maintained for corrosive storage tanks they would include:

1. Drawings and structural outlines
2. A listing of materials used for construction.
3. Welding certifications
4. Manufacturers warrantee
5. A listing of tests performed during and after construction.

6. Criteria required for specific mooring and mounting of the tank.
7. Periodic maintenance records.
8. Inspection history
9. Repairs, modifications, and changes made to the original design.

Note: National and international standards for corrosive tanks indicate that large tanks are to be placed in a location having secondary containment. EPA has set standards for above ground and below ground storage tanks. Anyone involved in building or installing tanks for corrosive substances should be knowledgeable about all federal, state, and local standards relating to storage of corrosive materials.

EMERGENCY PLANNING AND SPILL RESPONSE FOR CORROSIVE SUBSTANCES

A common response to controlling spills and releases of corrosive substances involves neutralization. Please be aware that if toxic or other inherent hazardous waste properties exist within a corrosive substance, neutralization may not keep spilled corrosive substances from requiring hazardous waste treatment. By all means, any controlled substance spills may require reporting to federal, state, and local regulatory authorities. The following are examples of emergency response procedures that may be used in controlling corrosive substance spills and releases to the environment.

Sample Guide for: Sodium Hydroxide, Caustic Soda (50%)

Storage: Bulk and 25% Totes
CAS: 1310-73-2
Hazard: Strong Alkali Corrosive
IDLH: 10 mg/cubic meter
MW: 40
PEL: 2 mg/ cubic meter
Flash Point: NA
DOT Hazard Class: 8, Corrosive Alkali, PGII
ERG Info: UN1823 Solid, UN1824 liquid ERG# 154

Reactive with: water, acids, flammable liquids, organic halogens, metals especially aluminum, tin, and zinc. Corrosive to most metals and also reacts with nitro methane.

Low volatility, unless involved in fire or high temperature situations. When reacting with combustible materials it can cause enough heat to ignite the combustible materials.

Personal Protection:

Level B if involved in high heat or reactive situations.

Level C at ambient temperatures and if air contaminant levels are within limitations of air purified respiratory protection.

Additional Considerations:

- Prevent contact with reactive substances
- Ensure PPE is alkali corrosive resistant
- Pump transfer spilled liquids to alkali resistant sealed containers
- Scoop and sweep spilled solids to alkali resistant sealed containers
- Neutralize residual and contaminated materials with vinegar or diluted weak, acetic acid.
- Absorb and dispose as hazardous waste if not neutralized or if contaminated with RCRA regulated materials.
- Significant releases to the environment require reporting to city, state, and federal regulatory agencies. RQ = 1000 lbs

Sample Guide for: Sulfuric Acid (93%)

Storage: Bulk
CAS: 7664-93-9
Hazard: Strong Acid Corrosive and Toxic, serious burns and destroys human tissues
IDLH: 15 mg / cubic meter
MW: 98.1
PEL: 1 mg / cubic meter
Flash Point: NA
DOT Hazard Class: 8, Corrosive, 6.1 Toxic, PG I
ERG Info: UN1831, ERG: #137

Reactive with: water, organic materials, chlorates, carbides, fulminates, powdered metals, corrosive to most metals, and generates enough heat to ignite combustible materials.

Personal Protection:
Level A Protection: Total encapsulation with corrosion protective plastics and rubber materials for responding to reactive spills that generate fumes, vapors, mists and other air contaminants.
Level B with Flash Protection: Total coverage protection with corrosive protective plastics and rubber materials and covered with an outer flash protective suit. For responding to reactive spills that have potentials to cause fire.

Additional Considerations:

- Prevent contact with reactive substances.
- Do not enter spilled liquid and avoid contact.
- Have fire suppression and appropriate fire fighting equipment near by.
- Pump transfer spilled liquids to acid resistant sealed containers.
- Neutralize residual and contaminated materials with soda ash.
- Absorb non-neutralized acids and materials with non-combustible, non-reactive absorbents and seal in acid resistant containers for disposal as corrosive and toxic hazardous wastes.
- Significant releases require reporting to city, state, and federal regulatory agencies. RQ = 1000 lbs

Acids, Anhydrides and Alkali

Safety Considerations

As previously mentioned, these materials will dissolve human flesh and metals. When metals are involved with corrosive substance exposure, flammable hydrogen gas is liberated. Along with damaging human, animal and plant tissues, corrosives are usually toxic as well. Many corrosive liquids are volatile, such that they rapidly enter into the air and surrounding atmosphere at ambient temperatures. Nitric acid and hydrochloric acid will liberate vapors into the air that can become dangerously concentrated.

Anhydrides have similar properties as acids. Usually, when they react with water they produce acid. In this regard, the chemical acetic anhydride when mixed with water will produce acetic acid.

As previously mentioned in Chapter 3, acids are neutralized using weak base materials - such as sodium carbonate. The end result of acid neutralization is formation of salt and water. Again, it is stressed that concentrated acids and alkalis should not be mixed together for the purpose of neutralization. Only neutralize concentrated acids with a weak alkali and only neutralize concentrated alkali with weak acids. The process of neutralization will yield hazardous vapors and increased temperatures (exotherm). Personnel performing neutralization must dress for the occasion, in corrosion resistant personal protective equipment.

Health Considerations

Acid and alkali chemicals burn human tissues. Amounts of exposure, coupled with concentrations of the corrosive substances yield health issues that range from irritation to second and third degree burns. As previously mentioned, corrosive damage to the eyes usually cannot be reversed. As off the wall as this may seem, teeth are also affected by corrosive vapors and fumes. Teeth can be damaged by hydrochloric acid fumes in concentrations as low as 10 parts per million.

Contamination of skin and clothing by corrosive dusts present health considerations that are often overlooked. As the tissues become moist, the dry corrosives begin to react and cause destruction. Caustic soda is known for its ability to cause latent burns. By the way, did you know that ammonia is considered to be a corrosive alkali chemical?

Workers in glass and electronics industries have seen serious effects when using chemicals that produce corrosive fluorides. In some situations, workers have witnessed extreme pain and tissue trauma from acid fluorides, such as HF or hydrofluoric acid. In this regard, HF causes the damaged skin to harden. The hardened surface tissues will cover the on-going necrotizing destruction that progresses deeply into the flesh. Victims often undergo painful treatments that eventually chemically bind and stop the corrosive tissue destruction.

Organic anhydrides will cause human tissues to become irritated and often sensitized. In this regard, exposure victims become plagued with dermatitis and in rare circumstances vexed with other health problems. Fortunately, most tissue burning activity from corrosive substances can be stopped by flooding with copious amounts of water for extended periods of time—if the exposure is noticed.

First Aid for Corrosive Exposures

The best time to plan for hazardous exposures is before you become exposed. As we have discussed in previous chapters, OSHA Chemical Hazard Communication requirements shall prevail. Anyone that works with hazardous substances, which includes corrosives, are expected to know and understand all properties of materials that they may become exposed. Also, they must know first aid procedures to perform when exposed. We must be familiar with information found in material safety and data sheets for every substance that we deal with. The U.S. Department of Health and Human Services has several publications, especially within the NIOSH Data Bases that are helpful in this regard. A most recent computer disk publication, DHHS (NIOSH) publication No. 2004-103 is a useful "no cost" resource to be used to encourage safe chemical work practices.

First Aid SOPs for Corrosive Exposure

- Move exposed victims to fresh air
- If not breathing and no heartbeat, perform CPR (cardio-pulmonary resuscitation)
- If possible place in shower
- Remove all clothing that was exposed to corrosive chemicals
- Flood exposed skin and eyes with copious amounts of fresh water
- To prevent vomiting, unconscious victims should be positioned on their side
- If corrosive substances have entered the mouth, rinse the mouth with water
- If corrosive substance has been swallowed, water may be given to drink in order to dilute concentration of corrosive in the stomach. Victim must receive immediate emergency medical treatment.
- While awaiting emergency medical treatment, keep victim warm and at rest

---

TYPICAL NEUTRALIZATON MATERIALS

To Neutralize Concentrated Acids
- o Sodium Bi Carbonate (Baking Soda)
- o Sodium Carbonate (Soda Ash)

To Neutralize Concentrated Alkali
- o Vinegar (dilute acetic acid)
- o Citric Acid

Figure 6.8 Typical neutralization materials.

TABLE 6.1 Corrosion resistant materials—resins used for making corrosion resistant equipment. *Reference: Corrosion Tables, Jernkontoret, Stockholm 1979.*

Resin	Maximum temperature (°C)	To be used in:
PVDF (polyvinylidene fluoride)	90	Storage tanks and pipes for sulfuric acid, nitric acid, hydrogen chloride, hydrogen fluoride
	85	Vaporizers
	70	Drying tower for chlorine
	70	Scrubbers and pipes for Sulfuric acid, phosphoric acid, hydrogen chloride, hydrogen fluoride
	100	Dilution tank for sulfuric acid
FEP (Teflon, fluorinated ethylene propylene)	95	Electrolytic bath
	120	Pipes
	90	Equipment for sulfuric acid regeneration
	90	Tanks containing chlorine
	100	Pipes for chlorine
LPE (linear high density polyethylene)	Room temperature	Laboratory bottles
PP (polypropylene)	80	Pipes for sulfuric acid, hydrogen chloride, hydrogen fluoride
	80	Entrainment separator for sulfuric acid, phosphoric acid, hydrogen fluoride
	70	
	70	Storage tanks for sulfuric acid, hydrogen chloride
	70	Gas pipes for sulfuric acid, sulfuric dioxide, sulfuric trioxide
	60	Scrubbers of sulfuric acid, phosphoric acid
		Blowers, fans
PVC (polyvinyl chloride, rigid)	50	Surface treatment baths and pipes
	50	Entrainment separator for sulfuric acid, chromic acid
	50	Pipes for sodium chlorate
	30	Storage tanks for hydrogen chloride
	40	Blowers, fans

TABLE 6.2 Corrosive effects on construction metals. Corrosion resistance assessed at 20° C if no other temperature indicated. *Reference: Corrosion Tables, Jernkontoret, Stockholm 1979.*

Substance	Aqueous solution concentration	Steel Fe 37 Fe 44 Fe 52	Austenitic stainless steel Cr 18%, Ni 9%	Austenitic stainless steel Cr 17%,Ni 12%, Mo 2.5%	Austenitic stainless steel Cr 20%, Ni 25%, Mo 4.5%, Cu 1.5%
Acetic acid	10%	+/-	+	+	+
	50%	-	+	+	+
	100%	-	+	+	+
Ammonia	all	+	+	+	+
Formic acid	5%	-	+	+	+
	50%	-	+	+	+
	100%	+/-	+	+	+
Hydrochloric acid	1%	-	+/-	+	+
	5%	-	-	-	-
	37%	-	-	-	-
Hydrofluoric acid	10%	-	-	-	-
	100%	+	+/-	+/-	+/-
Nitric acid	5%	-	+	+	+
	50%	-	+	+	+
	90%	-	+	+	+
Phosphoric acid	5%	-	+	+	+
	50%	-	+	+	+
	85%	-	+	+	+
Sodium hydroxide	10%	+	+	+	+
	50% (60°C)	+/-	+	+	+
Sulfuric acid	1%	-	+	+	+
	50%	-	-	-	+
	96%	+	+	+	+

Symbol Corrosion resistance   Corrosion speed
+   resistant   < 0.1 mm/year
+/-   not resistant, may be used only under specified conditions   0.1 -1.0 mm/year
-          not resistant, important corrosion effect, should not be used   > 1.0 mm/year

TABLE 6.3 Corrosive substance effects on concrete. *Reference: Corrosion Tables, Jernkontoret, Stockholm 1979.*

Substance	Solution	Effect on concrete
Acetic acid	10%	Slow decomposing effect
Acetic acid	30%	Slow decomposing effect
Acetic anhydride	concentrated	Slow decomposing effect
Ammonium hydroxide		No
Arsenic acid		No
Barium hydroxide		No
Boric acid		Negligible
Calcium hydroxide		No
Carbon bi-sulfide		No, reacts with water and humidity to produce sulfuric acid which has highly corrosive effect
Carbonic acid		Slow decomposing effect, also on steel
Chromic acid	5%	May be corrosive to steel
Chromic acid	50%	May be corrosive to steel
Cotton seed oil		Decomposing effect in presence of air
Formic acid	10%	Slow decomposing effect
Formic acid	30%	Slow decomposing effect
Hydrogen chloride	10%	Quick corrosive effect, also on steel
Hydrogen chloride	37%	Quick corrosive effect, also on steel
Hydrogen fluoride	10%	Quick corrosive effect, also on steel
Hydrogen fluoride	75%	Quick corrosive effect, also on steel
Nitric acid	2%	Quick corrosive effect
Nitric acid	10%	Quick corrosive effect
Nitric acid	30%	Quick corrosive effect
Ortho-phosphoric acid	10%	Slow decomposing effect
Ortho-phosphoric acid	85%	Slow decomposing effect
Oxalic acid		No, anticorrosive effect on tanks
Perchloric acid	10%	Corrosive
Potassium hydroxide	5%	No
Potassium hydroxide	25%	Corrosive
Potassium hydroxide	95%	Corrosive
Sodium hydroxide	1%	No
Sodium hydroxide	10%	No
Sodium hydroxide	20%	Corrosive
Sodium hydroxide	40%	Corrosive
Sulfuric acid	10%	Destroys quickly
Sulfuric acid	50%	Destroys quickly
Sulfuric acid	80%	Destroys quickly
Sulfuric acid	93%	Corrosive
Sulfuric acid	concentrated	Corrosive
Sulfurous acid		Quick corrosive effect
Tartaric acid solution		No

TABLE 6.4 Substances that are non-compatible with corrosives. *Reference: Corrosion Tables, Jernkontoret, Stockholm 1979. (continued on next page)*

Compound	is incompatible with:
Acetic acid	chromic acid, nitric acid, hydroxyl compounds, ethylene glycol, perchloric acid, peroxides, permanganates
Acetylene	chlorine, bromine, fluorine, copper, silver, mercury
Acetone	concentrated nitric acid and sulfuric acid mixtures
Ammonia (anhydrous)	mercury, chlorine, bromine, iodine, calcium hypochlorite, hydrofluoric acid (anhydrous)
Ammonium nitrate	acids, powdered metals, flammable liquids, chlorates, nitrites, sulfur, finely divided organic or combustible materials
Aniline	nitric acid form mixture which is auto-ignites
Azides	acids
Bromine, chlorine	ammonia, acetylene, butadiene, butane, methane, petroleum gases, hydrogen, sodium carbide, benzene, finely divided metals, turpentine
Calcium oxide	reacts with water producing heat
Carbides, nitrides, sulfides	produce toxic and flammable gases (such as ammonia, hydrogen sulfide, acetylene) in contact with acids
Chromic acid	acetic acid, naphtalene, camphor, glycerol, alcohols, flammables
Cyanides	form highly toxic hydrogen cyanide in contact with acids
Fluorine	reacts with a wide range of substances and construction materials, practically with everything
Hydrogen peroxide	copper, chromium, iron, most metals and their salts, alcohols, acetone, aniline, nitromethane, organic or combustible materials
Hypochlorites	decompose at room temperature when in contact with acids, producing corrosive and toxic gases
Iodine	acetylene ammonia (aqueous or anhydrous), hydrogen
Nitrates	sulfuric acid

TABLE 6.4 Substances that are non-compatible with corrosives. *Reference: Corrosion Tables, Jernkontoret, Stockholm 1979. (continued from previous page)*

Nitric acid	acetic acid, aniline, chromic acid, cyanides, sulfides, flammable solids, liquids or gases, copper, brass, heavy metals
Nitroparaffins	alkalis, such as sodium hydroxide and potassium hydroxide, amines
Organic peroxides	Are thermally unstable substances which react in contact with strong acids and alkalis, and may undergo self-accelerated decomposition, with danger of fire and explosion.
Oxygen	oils, grease, hydrogen; reacts with combustible material: solids, liquids and gases
Peroxides, organic	acids, avoid friction, store cold
Potassium chlorate, sodium chlorate	in contact with strong acids (such as sulfuric acid) gives off toxic chlorine dioxide
Potassium permanganate	glycerol, ethylene glycol, benzaldehyde, sulfuric acid
Silver	acetylene, oxalic acid, tartaric acid, ammonium compounds, fulminic acid
Sodium hydroxide, potassium hydroxide	In contact with aluminum zinc or galvanized metals may give off flammable gases: hydrogen and dichloroacetylene. (Dichloroacetylene forms explosive mixtures with trichloroethylene.)
Sodium nitrite	ammonium nitrate and other ammonium salts
Sodium peroxide	ethyl or methyl alcohol, acetic anhydride or glacial acetic acid, benzaldehyde, carbon disulphide, glycerin, ethylene glycol, ethyl acetate, methyl acetate, furfural
Sulfuric acid	potassium chlorate, potassium perchlorate, potassium permanganate and similar compounds of sodium, magnesium and lithium

Chapter 7

IGNITABLE SUBSTANCES

CHARACTERISTICS OF IGNITABLE HAZARDOUS SUBSTANCES

As described in Chapter 3, ignitable hazardous substances are materials that will readily catch fire, explode, or play a major roll in uncontrolled burning reactions. They may be usable products or hazardous wastes, which are ignitable materials that no longer have value. The physical characteristics of ignitable hazardous substances include materials that are:

- Flammable liquids
- Combustible liquids
- Ignitable gases
- Ignitable solids

FLAMMABLE AND COMBUSTIBLE LIQUIDS

Flammable and combustible liquids will readily ignite and burn. Each substance is characterized by its flashpoint or "temperature that it will become air" and as described in chapter 3 and in the glossary of this text, will explode or burn when the airborne vapors contact an ignition source.

Please note that comprehensive information related to physical and health properties of many substances has been assembled by the United States Department of

Health and Human Services (NIOSH) in a comprehensive computer disk and it is available at little to no cost. Contact 1-800-35-NIOSH and request a copy of DHHS (NIOSH) Publication No. 2004-103.

Ignitable liquids are usually considered flammable if they have a flashpoint at or below 100 degrees Fahrenheit (37.8 degrees C). Ignitable liquids are considered combustible if they have a flashpoint above 100 degrees Fahrenheit and below 200 degrees Fahrenheit (93.3 degrees C). Please note as described in Chapter 3, that any ignitable liquid having a flashpoint of 140 degrees or less is considered flammable by the United States Department of Transportation, International Civil Aviation Organization, and U.S. EPA hazardous waste standards in the Resource Conservation and Recovery Act (RCRA). In this regard, any ignitable liquid having a flashpoint at or below 140 degrees Fahrenheit, are considered flammable in this text as well.

Ignitable liquids are commonly found in every environment. This is especially true since certain special interest groups encouraged EPA and Congress to accept somewhat questionable scientific data to establish laws that forbid use of non-ignitable, low toxicity, chloro-fluoro carbons in aerosol products. Please note that we are now using dangerously flammable products that yield significant human health hazards. Look in your kitchen and bathroom cabinets, walk through the grocery store, and read the constituents listed on product labels for most spray preparations. Reference the listed constituents in the above mentioned DHHS (NIOSH) chemical safety data. Then ask the question, should we as a world community be concerned? The answer should be quite obvious. Individual safety, public safety, and human health should be of prime concern now and in the future.

OTHER CONSIDERATIONS FOR IGNITABLE LIQUIDS

Also covered in Chapter 3 are the physical properties of all substances that make them unique such as volatility, vapor density, incompatibility and other concerns. We need to further discuss these considerations in terms of ignitable liquids safety.

Periodically we hear about people that become burned to death as they are filling gasoline in their lawn mower. How about fire department and industrial haz-mat teams that respond to manage ignitable liquid spills and were never trained to deal with specific properties of ignitable releases? Some personnel do not follow product specific standard operating procedures, some of their employers cannot afford to use flash protection equipment; unfortunately we hear the news about them also.

Most ignitable chemical products yield more than one hazard and along with flash protection, other hazardous properties need to be considered. A good source for characterizing multiple chemical hazards is using the International Safety Card system as shown on the next page.

IGNITABLE LIQUID HAZARDS CONSIDERATIONS

Review the following International Safety Card that is provided by the Centers for Disease, and National Institute of Occupational Safety and Health (NIOSH), and note the all the considerations that are listed for acetaldehyde, reference DHHS (NIOSH) publication No. 2004-103.

Note that the following considerations are addressed and should be noted by anyone involved with products that are hazardous. Also note that sometimes material safety data sheets (MSDS) may not list all significant data that workers and emergency personnel may need to know for every situation.

International Chemical Safety Cards (ICSC) Considerations

- Acute and Chronic Hazards
- Accident, Incident Prevention
- First Aid and Fire Fighting
- Spillage Disposal
- Product Storage
- Packaging and Labeling
- Physical Appearance
- Physical & Chemical Dangers
- Occupational Limitations
- Routes of Exposure, Effects of Exposure
- Inhalation Risks
- Physical Properties
- Environmental Data
- Emergency Data

As an example, see Figure 7.1, International Chemical Safety Cards beginning on the next page.

IGNITABLE LIQUID SPECIFICS

Prescribed safe handling of ignitable liquids is based upon specific physical characterizations of combustible and flammable liquids.

Combustible Liquids

Liquids that will burn, that have flash points at or above 100 degrees Fahrenheit (37.8 degrees C) are considered combustible. They are divided into two classifications:

International Chemical Safety Cards

ACETALDEHYDE   **ICSC:**

0009

aldehyde
Molecular mass: 44.1

Acetic aldehyde
Ethanal
Ethyl
C_2H_4O / CH_3CHO

ICSC #   0009
CAS #   75-07-0
RTECS #   AB1925000
UN #   1089
EC #   605-003-00-6

TYPES OF HAZARD/ EXPOSURE	ACUTE HAZARDS/ SYMPTOMS	PREVENTION	FIRST AID/ FIRE FIGHTING
FIRE	Extremely flammable.	NO open flames, NO sparks, and NO smoking. NO contact with hot surfaces.	Powder, alcohol-resistant foam, water in large amounts, carbon dioxide.
EXPLOSION	Vapour/air mixtures are explosive.	Closed system, ventilation, explosion-proof electrical equipment and lighting. Do NOT use compressed air for filling, discharging, or handling. Use non-sparking handtools.	In case of fire: keep drums, etc., cool by spraying with water.
EXPOSURE		PREVENT GENERATION OF MISTS! STRICT HYGIENE!	IN ALL CASES CONSULT A DOCTOR!
•INHALATION	Cough. Drowsiness. Shortness of breath. Unconsciousness. Symptoms may be delayed (see Notes).	Ventilation. Local exhaust or breathing protection.	Fresh air, rest. Half-upright position. Refer for medical attention.
•SKIN	Redness. Burning sensation. Pain.	Protective gloves. Protective clothing.	Remove contaminated clothes. Rinse and then wash skin with

Figure 7.1   International Chemical Safety Card for Acetaldehyde (DHHS/NIOSH).
(continued on next page)

			water and soap. Refer for medical attention.
•EYES	Redness. Pain. Blurred vision.	Safety goggles, or eye protection in combination with breathing protection.	First rinse with plenty of water for several minutes (remove contact lenses if easily possible), then take to a doctor.
•INGESTION	Burning sensation. Diarrhea. Dizziness. Nausea. Vomiting. Further see Inhalation).	Do not eat, drink, or smoke during work.	Rinse mouth. Give plenty of water to drink. Refer for medical attention.

SPILLAGE DISPOSAL	STORAGE	PACKAGING & LABELLING
Evacuate danger area! Eliminate ignition sources! Collect leaking and spilled liquid in sealable containers as far as possible. Absorb remaining liquid in sand or inert absorbent and remove to safe place. Do NOT wash away into sewer. Do NOT absorb in saw-dust or other combustible absorbents. Remove vapour with fine water spray. (Extra personal protection: self-contained breathing apparatus).	Fireproof. Separated from incompatible substances (see Chemical Dangers). Cooled. Keep in the dark. Store only if stabilized.	Unbreakable packaging; put breakable packaging into closed unbreakable container. F+ symbol Xn symbol R: 12-36/37-40 S: 2-16-33-36/37 UN Hazard Class: 3 UN Packing Group: I

SEE IMPORTANT INFORMATION ON BACK

ICSC: 0009	Prepared in the context of cooperation between the International Programme on Chemical Safety & the Commission of the European Communities (C) IPCS CEC 2002. No modifications to the International version have been made except to add the OSHA PELs, NIOSH RELs and NIOSH IDLH values.

Figure 7.1   International Chemical Safety Card for Acetaldehyde (DHHS/NIOSH).
(continued from previous page)

PHYSICAL STATE; APPEARANCE:
GAS OR COLOURLESS LIQUID, WITH PUNGENT ODOUR.

PHYSICAL DANGERS:
The vapour is heavier than air and may travel along the ground; distant ignition possible.

CHEMICAL DANGERS:
The substance can form explosive peroxides in contact with air. The substance may polymerize under the influence of acids, alkaline materials, such as sodium hydroxide, in the presence of trace metals (iron) with fire or explosion hazard. The substance is a strong reducing agent and reacts violently with oxidants. Reacts violently with various organic substances, halogens, sulfuric acid and amines, causing fire and explosion hazard.

OCCUPATIONAL EXPOSURE LIMITS:
TLV: 100 ppm; 180 mg/m^3 as TWA (ACGIH 1991-1992). TLV (as STEL): 150 ppm; 270 mg/m^3 (ACGIH 1991-1992). OSHA PEL: TWA 200 ppm (360 mg/m^3) NIOSH REL: Ca See Appendix A See Appendix C (Aldehydes) NIOSH IDLH: Potential occupational carcinogen 2000 ppm

ROUTES OF EXPOSURE:
The substance can be absorbed into the body by inhalation and by ingestion.

INHALATION RISK:
A harmful contamination of the air can be reached very quickly on evaporation of this substance at 20°C.

EFFECTS OF SHORT-TERM EXPOSURE:
The substance irritates the eyes, the skin and the respiratory tract. Inhalation of the vapour may cause lung oedema (see Notes). The substance may cause effects on the central nervous system, resulting in lowering of consciousness. The effects may be delayed.

EFFECTS OF LONG-TERM OR REPEATED EXPOSURE:
Repeated or prolonged contact with skin may cause dermatitis. The substance may have effects on the central nervous system, respiratory tract and kidneys, resulting in chronic alcohol-like intoxication. This substance is possibly carcinogenic to humans.

I M P O R T A N T   D A T A

Figure 7.1 International Chemical Safety Card for Acetaldehyde (DHHS/NIOSH). *(continued from previous page)*

| PHYSICAL PROPERTIES | Boiling point: 21°C
Melting point: -123°C
Relative density (water = 1): 0.78
Solubility in water: miscible
Vapour pressure, kPa at 20°C: 99 | Relative vapour density (air = 1): 1.5
Flash point: -39°C
Auto-ignition temperature: 185°C
Explosive limits, vol% in air: 4-57
Octanol/water partition coefficient as log Pow: 0.43 |
| ENVIRONMENTAL DATA | | |

N O T E S

| IMPORTANT LEGAL NOTICE: | Neither NIOSH, the CEC or the IPCS nor any person acting on behalf of NIOSH, the CEC or the IPCS is responsible for the use which might be made of this information. This card contains the collective views of the IPCS Peer Review Committee and may not reflect in all cases all the detailed requirements included in national legislation on the subject. The user should verify compliance of the cards with the relevant legislation in the country of use. The only modification made to produce the U.S. version is inclusion of the OSHA PELs, NIOSH RELs and NIOSH IDLH values. |

Figure 7.1 International Chemical Safety Card for Acetaldehyde (DHHS/NIOSH).
(continued from previous page)

1. Class II Liquids:Substances with flashpoints at or above 100 degrees Fahrenheit but less than 140 degrees Fahrenheit (60 degrees C). The concentration of Class II combustible liquids contained in a given substance is at least 99%.
2. Class III Liquids:Substances with flashpoints at or above 140 degrees Fahrenheit and that meet the following 2 classifications:
   a. Class IIIA Combustible liquid with flashpoints at or above 140 degrees and below 200 degrees Fahrenheit with combustible concentrations of at least 99%.
   b. Class IIIB Combustible liquid with flashpoints above 200 degrees Fahrenheit (93.3 degrees C).

Flammable Liquids

Liquids that will burn, that have flashpoints below 100 degrees Fahrenheit are divided into three classifications:

1. Class 1A includes liquids with flashpoint temperatures below 73 degrees Fahrenheit, and boils below 100 degrees F.
2. Class 1B includes liquids with flashpoint temperatures below 73 degrees F, and boils above 100 degrees F.
3. Class 1C includes liquids with flashpoint temperatures at or above 73 degrees F, and below 100 degrees F.

IGNITABLE LIQUID STORAGE

Ignitable liquids stored in drums and other containers are usually referred as individual capacity storage at quantities less than 60 gallons and portable capacity at quantities less than 660 gallons. Portable tanks are closed containers that contain more than 60 gallons and are not intended for use at "fixed" installations, such as chemical factories, distilleries, refineries etc...

Storage Specifics

Approved ignitable liquid storage containers would meet criteria described in the Department Transportation standards 49 CFR 173 - 178 and the National Fire Protection Association, NFPA 30. Portable tanks are required to have emergency venting. Top mounted vents shall be capable of limiting internal pressure when exposed to extremes of heat. Internal pressure should be limited to the greater of 10 psig or 30% of tank bursting pressure. Portable tanks should also have at least one pressure activated vent that would be set to open at 5 psig (minimum). Fusible vents are sometimes used. Fusible vents require activation at temperatures of 300 degrees Fahrenheit or less.

Should plastic containers be used to store ignitable liquids?

Sizes and shapes of containers used for storing ignitable liquids are usually determined by physical properties of substances to be stored. You may have noticed that sources for gasoline containers often sell plastic containers. Is this legal? The author believes that ignitable liquids should be stored in containers that can be grounded to prevent sparks from static electricity. Plastic is not electrically conductive.

Please know that exceptions and exemptions exist that allows use of plastic containers for ignitable liquids, each having limitations and specifications that must be complied with to allow use. The following standards describe use of plastic containers for storing ignitable liquids:

- ANSI/ASTM D 3435-80: Plastic Containers (Jerry Cans) for Petroleum Products.
- ASTM F 852-86: Standard for Portable Gasoline Containers for Consumer use.
- ASTM F976-86, Standard for Portable Kerosene Containers for Consumer use.
- ANSI/UL 1313-83, Non-metallic Safety Cans for Petroleum Products.

Criteria for Ignitable Liquid Storage Cabinets

Quantities of ignitable liquids that may be kept in storage cabinets are as follows:

- Class I and or Class II ignitable liquids should not be stored in quantities greater than 60 gallons.
- Class III ignitable liquids should not be stored in quantities greater than 120 gallons.
- Metal storage cabinets are recommended for the purpose of grounding.
- Ignitable liquid storage cabinets should keep internal temperatures from exceeding 325 degrees Fahrenheit when subjected to a standardized 10 minute fire test.
- All attachments, joints, and seams must be able to remain tight and the door must be able to remain securely fastened during a standardized 10 minute fire test.
- The construction of ignitable liquid storage cabinets requires at least Number 18 gage sheet metal, formed as double walls that have a 1 ½ inch air space.
- Cabinet doors require 3 point locking mechanisms and the door sill shall be at least 2 inches above the cabinet bottom.
- Ignitable liquid storage cabinets require marking with contrasting printing that states "Flammable - Keep Fire Away."

Ignitable Substance "Authorized" Storage Rooms

The NFPA Standard 251-1969, "Standard Methods of Fire Tests of Building Construction, and Materials," describes requirements for the construction of inside ignitable substance storage rooms.

The following should be considered:

- Openings to adjoining rooms or spaces: Requires non-combustible, raised, liquid tight sills, or ramps that are at least 4 inches in height, or the storage room floor should be at least 4 inches below the surrounding floor.
- Openings require approved, self closing fire doors.
- The walls in the room shall be liquid tight at points where the walls connect with the floor.
- In place of a sill or ramp: An open grated trench may be installed inside the room - provided that any ignitable liquid spillage would drain from the trench to a collector tank that is located in a safe location.

STORAGE ROOM RATING AND CAPACITY

Figure 7.2 indicates considerations for ignitable liquid storage within rooms.

Fire protection - The following should be considered based upon materials stored: Use of water sprinkler system, water spray, carbon dioxide, other systems, and overflow recovery.

FIRE PROTECTION PROVIDED	FIRE RESISTANCE	MAXIMUM FLOOR AREA	ALLOWABLE QUANTITIES
Yes or No	# hours	Square Feet	Gallons/ Square Foot
Yes	2 hrs	500	10
No	2 hrs	500	5
Yes	1 hr	150	4
No	1 hr	150	2

Figure 7.2 Fire resistance.

Note: See additional considerations for ignitable storage rooms on following pages for:

- Electrical Wiring
- Ventilation and Air Quality
- Storage

Electrical Wiring

Electrical wiring and equipment located in storage rooms that contains ignitable liquids, is required to meet standards described in OSHA Standard, Subpart S, 29 CFR 1910.307 which deals with electrical installations for hazardous classified locations. The overall summary of the OSHA directive is that equipment, wiring methods, and installations of equipment in locations that are designated hazardous must be intrinsically safe and approved. Electrical systems and related components in ignitable substance storage rooms shall be explosion proof.

Note: Ignitable liquid storage containers and cabinets must be grounded to prevent discharge of static electricity. When ignitable liquids are transferred from storage containers to other containers, all transfer system components must be grounded.

Ventilation and Air Quality

Storage rooms that contain ignitable liquids must be ventilated. Venting is critical to insure that air concentrations do not contain explosive or flammable quantities of ignitable vapors or gases. Arrangement of ventilation is extremely important, knowledge of vapor densities for all substances must be considered. Airflow should be directed so that vapors and gases are moved out of the room and fresh air shall be maintained at a constant flow.

According to OSHA directive 29 CFR 1910.106 (d) (4) (iv), hazardous storage rooms shall contain either gravity or mechanical exhaust. Air within the storage room is required to completely change over at least six (6) times per hour. When mechanical exhaust is used, it should be controlled by a switch that is located outside of the storage room. Electrical components within the room should be controlled by the switch located outside the room. If Class 1 Flammable Liquids are dispensed within the storage room, a pilot light is required to be installed adjacent to the control switch located outside of the room. When gravity ventilation is used, fresh air intake and exhaust outlet from the storage room is required to be directed outside of the building.

Solvent Room Storage

Storage within ignitable liquid storage rooms must be arranged so that isles or walkways are clear for movement and they should be at least 3 feet wide. Containers that hold 30 gallons or more of ignitable liquid shall not be stacked upon each other. Dispensing of ignitable liquids requires use of approved pumps or self closing faucets.

Storage of ignitable liquids inside buildings must be arranged so they do not limit use of exits, stairways, or any area used for safe egress of personnel.

Ignitable liquids stored in buildings having office occupancies are basically forbidden except for quantities required to perform maintenance and to operate equipment. Storage in this regard, requires tightly closed containers that shall be stored in authorized cabinets or in rooms that do not open toward building occupants; or do not allow direct access to occupied areas of the building.

Damaged Containers or Leakage

For ignitable liquid spillage, refer to specific standard operating procedures required for responding to emergency actions [HAZWOPER] listed in Chapter 5 of this text. When containers are damaged or leaking, they must be removed to a safe location outside the building or to an authorized ignitable liquid storage room to transfer the liquid to an authorized, undamaged container.

General Purpose Public Warehouses

The OSHA Standard 29 CFR 1910.106 (d)(5)(v) directs proper storage of ignitable liquids in warehouse buildings that are cut off by standard firewalls.

Tables H-14 and H-15 within the standard list specific criteria for indoor container and portable tank storage. Incompatible materials that increase fire potential or create fire hazards are not permitted within the storage areas.

Examples non-permitted materials include:

- Reactive materials
- Oxidizers
- Corrosives

Other inside storage considerations

- Egress: Entry and exit of buildings that contain ignitable hazards must not be restricted. All entry ways, exits, stairways, and areas used for safe egress of people cannot be blocked or restricted.
- Office Buildings: Ignitable substance storage is prohibited except for materials used for maintenance and for operating equipment, such as emergency power generation and others. Containers must be kept closed and stored in authorized flammable storage cabinets. Reactive substances are not allowed in flammable and combustible material storage areas.

IGNITABLE SUBSTANCE SAFETY

Properties of Flammable and Combustible Liquids

When in open containers, spilled or released to the environment, flammable and combustible liquids will vaporize and blend into the air. As mentioned in Chapter 3 or this text, safe handling requires knowledge of the properties of volatility, flash point, vapor density, and burning range (LEL - UEL) for each ignitable substance. Remember, if we know where a given substance is when it becomes air, we are likely to properly control it by using ventilation and other forms of engineering controls. Danger: Do not ventilate by placing a non-explosion proof fan in a potentially flammable environment. Air testing should be performed using a combustible gas indicator (CGI) or other devices, to determine concentrations of combustible or explosive products in the air. *Remember it is dangerous to enter an ignitable atmosphere that is greater than 10% of the LEL.*

Storage Reminder

As previously mentioned, ignitable materials must be stored in containers and cabinets as prescribed by the United States Department of Transportation (DOT), the U.S. Occupational Safety and Health Administration (OSHA), the National Fire Protection Association (NFPA), the American National Standards Institute (ANSI) and the Uniform Fire Code.

Containers and Cabinets

- Keep containers closed, by sealing with a lid or device to prevent leakage or evaporation into the environment.
- Use approved fire safety containers that do not hold more than 5 gallons, and that has a spring closing lid with a flame arrester device built within the spout.
- Be sure to properly ground any container that is used to transfer ignitable substances, any container that ignitable substances are transferred to and any cabinet that holds ignitable substances.
- Ignitable substance storage cabinets should be designed to protect stored contents from external ignition sources.
- Storage cabinets should be marked as potential fire hazards.
- Out of storage quantities should be limited to "only enough required to do the work"…
- No more than 25 gallons of ignitable liquids total, should be used outside of storage cabinets, and they should be stored in authorized safety cans.

- Don't store incompatible substances with ignitable substances, that includes oxidizers - even though the tag is marked with a flame on it.
- In laboratories that require refrigerator storage; the refrigerators must be specially designed and manufactured for flammable liquid storage.

Safe Handling Reminder

To safely work with and around ignitable hazardous liquids, gases, or solids the following should be considered:

- Control potential ignition sources.
- Keep in approved storage containers and cabinets.
- Do not store near or with oxidizers
- Keep away from heat sources and out of direct sunlight.
- In flammable environments, use only spark resistant, explosion proof devices.
- Ground containers and transfer systems to prevent static electricity generation. Note: In some cases, cool dry environments may be more prone towards static electricity. Also, if non-electrically conductive containers are used - grounding contact should be made directly to the liquid and not the container.
- Don't allow people to smoke or use open flames near or around ignitable hazardous substances.
- Ensure that processes and related equipment do not produce ignition sources.
- Use safety containers approved for handling ignitable hazardous substances...
- Ventilate fumes, vapors, dusts, mists or other ignitable airborne ignitable hazardous substances. (Locate ventilation ports, air make up and exhaust, based upon the vapor densities of ignitable or explosive substances that are emitted into the air space).
- Use approved bottle carriers when transporting ignitable liquids in glass or plastic containers.

IGNITABLE SOLIDS

Substances such as alkali metals, magnesium powders and shavings, metal hydrides, some of the organo-metallic compounds and sulfur are examples of ignitable solids. Some ignitable solids react with water and will also react with typical dry chemical and carbon dioxide fire extinguishers. In fact, fires that involve lithium aluminum hydride will explode when exposed to carbon dioxide. Class D fire extinguishers are used for ignitable solid fires (some times sand may be used to smother an ignitable solid fire).

Catalytic Ignition

In laboratory conditions, some materials used as catalysts such as palladium, platinum oxide, Raney Nickel and others may become saturated with hydrogen. When this occurs, an explosion and or fire may occur. To prevent this from occurring, the lab technician may:

- Filter the catalyst
- Being careful to keep the filter cake from becoming dry
- Then, immediately placing the funnel and moist catalyst into a water bath.

Note: Purging gas such as argon or nitrogen may be used to inert the atmosphere around the catalyst.

Pyrophoric Substances

Pyrophoric substances can spontaneously ignite when exposed to normal air concentrations. The term pyrophoric is derived from the Greek language and means "fire bearing."

Some pyrophoric substances will react with water or high humidity, and rapidly ignite.

Examples of pyrophoric substances include organic-metallic materials such as compounds of alkyl lithium, alkyl zinc, alkyl magnesium (Grignard Reagents) and some finely divided metal powders.

Because of the reactive and hazardous nature of these substances, special training is required to safely handle and store them. Most commonly, pyrophoric substances are handled in inert or un-reactive environments that utilize special equipment and nitrogen or argon gas.

Fires involving pyrophoric substances usually require Class D fire extinguishers for control. Due to hazards related to fighting pyrophoric substance fires, personnel require hands on training in safe extinguishment of fires involving pyrophoric substances. Carbon dioxide extinguishers will cause the fire to become more intense.

Another Thought About Ignitable Solids

Most people do not think that metal dusts or particles will burn as vigorously as they do. Burning Aluminum and magnesium dusts have caused deaths and severe destruction in this regard. In situations that involves flammable metals such as lithium, sodium or potassium, or organic-metal compounds such as butyl-lithium, diethyl zinc and others, please remember that water, carbon dioxide and typical dry extinguishing agents can make the fires worse—learn to use safe Class D extinguishing techniques before fighting a fire that involves ignitable solids.

ABOUT DUSTS

Where do dusts come from? Dusts come from many sources. Dusts from food grade materials, and other non-hazardous sources can explode or cause fire. Dusts become suspended in air and remain evenly dispersed in many cases. Dusts are different than fumes, vapors or mists, they are composed of solid particles that collectively form large surface areas that one spark can initiate a chain reaction that violently releases energy.

From human health perspectives, dust particles can be drawn into your lungs. As mentioned with asbestos, lead and other substances that cause both acute and chronic health effects, respiratory protection and engineering controls are required.

How do we prevent explosive or health hazard concentrations of dusts to form?

- Control dust at the source using engineering controls, for example a local exhaust venting system.
- Operations that generate dust often use dust collection systems that dampen generated dusts with moisture and filter or collect the dispersed dusts.
- From health hazard perspectives dust concentrations may be lowered below explosive levels; however, the atmosphere remains a health hazard. In this case, personal protective equipment should include at minimum respiratory protection and eye protection.

ABOUT FUMES

Often fumes are regarded as vapors, dusts, smoke, or a mixture of each. Also in the case of welding operations, vaporized metals can become airborne. In this regard, fumes are very fine particles that are suspended in air. Fumes can be ignitable, toxic, corrosive, as well as demonstrate other hazards. An example is titanium tetrachloride that immediately reacts violently with water vapor in the air and forms dense white fumes. The following reaction shows what takes place:

$$TiC_{l4} + 2\ H_2O\ TiO_2(s) + 4\ HC_l(g)$$

In the reaction, titanium and oxygen form solid small white particles. Acidic HCl (hydrogen chloride) is an invisible and hygroscopic corrosive and toxic gas that immediately condenses with water vapor in the air to form suspended liquid particles. By inhaling fumes produced by this reaction, the victim will suffer burns to the lungs that can easily yield chemical pneumonia and other destructive long term health effects.

ABOUT VAPORS

Vapor also can be ignitable, toxic, corrosive, and produce negative health effects. Vapors are the result of the gas phase that a given material may exhibit. The critical temperature at which this effect occurs varies with each substance. A common example is the water vapor phase, also called gas phase of water. Vapors should not be confused with dusts or mists that are fine suspensions of liquids in the air. When a substance changes from a solid or liquid and becomes gaseous, the process is called vaporization.

SOLID, LIQUID, OR GAS

The distinction between each phase deals with the arrangement of atoms and eventual forming of molecules that characterizes each material:

- Solids: Demonstrate strong forces between each grouping of atoms or molecules. Each respective molecule has a close interaction with surrounding molecules to form tight spatial arrangements that resist change. In this regard, solids present an ordered arrangement of molecules that are referred to as crystalline. There are some solids that do not have a tight arrangement of order, they are called amorphous.
- Liquids: Substances that are in liquid form demonstrate weak interactions between atoms and molecules. Liquid molecules are free to move past one another. In this regard, increased temperature causes rapid movement.
- Gases: Substances that demonstrate small amounts of molecular interactions. Distances between molecules are much greater than in liquids and solids. Gases easily diffuse and bounce around inside a container.

ABOUT MIST

Mist, also called fog, is usually discussed when relating to water vapor. However, anyone that has removed an automobile gasoline tank cap during July or August may have noted moisture or liquid droplets that are forced by expansion from the gas tank. By the way, move away from the person that is smoking when removing a gasoline filler cap or during refueling - that person is trying to become a "crispy critter."

Sometimes the term "fog" is confused with vapor. Vapors are single molecules that exist in a gas phase, whereas fog or mists are millions of molecules that remain in the liquid phase. Most vapors can't be seen by the naked eye; usually fogs or mists can be seen. Mists can produce a wet film on most surfaces. Ignitable fogs or mists are often referred to as "Flammable Aerosols."

ABOUT IGNITABILITY AND EXPLOSIVES

As we have previously mentioned, gasoline readily forms an explosive mixture with air. If it didn't, we wouldn't be using gasoline engines in automobiles. The general definition for an explosive is a substance that is affected by heat, shock, friction, or other energies and causes immediate releases of energy and gas as by-products from decomposition and chemical reconfiguration. Dangerously, the chemical reconfiguration yields volumes of heat and gases much greater than was exhibited by the original or starting materials.

Explosives are usually characterized in two general classifications:

- High Explosives
- Low Explosives

High Explosives

High explosives will detonate or react to cause an energy release wave that moves quicker than the speed of sound (up to 9,000 meters per second). High explosives are characterized as Primary Explosives and Secondary Explosives:

- Primary Explosives include substances such as nitroglycerin, which can detonate from moving it around from place to place.
- Secondary Explosives include substances such as TNT (trinitrotoluene) or dynamite. To explode, they usually require fire, electrical impulse, blasting caps, or detonators.

Low Explosives

Low Explosives are usually characterized by combustion or deflagration, which is rapid burning without generating a wave of high pressure (usually less than 2,000 meters per second). For example, gunpowder and black powder are Low Explosives.

ENERGIES OF IGNITABILITY

Exothermic reactions are energy releases to the environment such that heat is generated. Endothermic reactions involve collecting heat from the environment. Thermoneutral indicates that heat is not released or collected.

DECOMPOSITION

Whenever a substance changes as a result of exposure to heat, electrolysis, chemical reaction, biological activity, as well as by other mechanisms, the end result is often called decomposition. An example of sugar decomposition is shown in Figure 7.3.

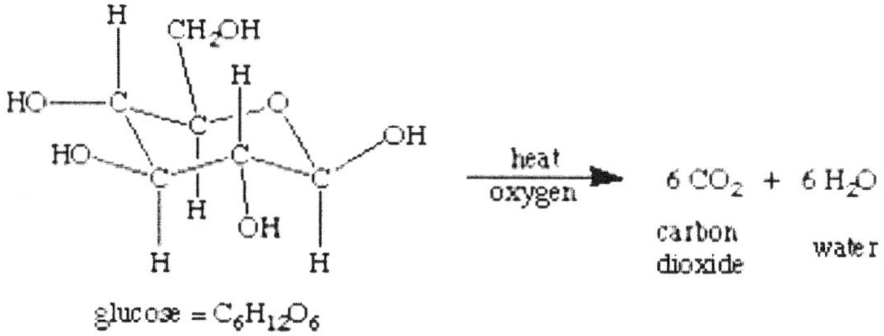

Figure 7.3 Decomposition of sugar.

As a result of burning or combustion, sugar (glucose) decomposes or reconfigures into carbon dioxide and water. Some "Mad Chemists" may call this reaction reconjugation of sugar due to heat decomposition. Unfortunately the presence of other substances that may be involved in the fire may reconfigure into monoxides, cyanides, chlorides, and other toxic stuff.

HAZARDOUS POLYMERIZATION

Fortunately during this dispensation time, we enjoy materials that are made from hydrocarbon chains, not just petroleum products or gasoline, but polymers. Just about everything we use involves polymers in some shape or form. Carbon units called monomers are joined together by a chemical reaction to form polymers. Some of us call this carbon-building process "tinker toy chemistry." Polymers are primarily characterized by their molecular mass or formula weight.

Unfortunately, some products can undergo hazardous polymerization during an emergency or whenever a polymer building reaction becomes out of control. This is termed "runaway polymerization" which often results in a catastrophic explosion. The product Ethylene, see Figure 7.4, is subject to this reaction. Ethylene (C_2H_4) is extremely volatile and flammable. In the presence of a catalyst, ethylene will form solid straight carbon chains as CH_2 called polyethylene, commonly used in plastic. A catalyst is a product not consumed in the reaction but used to control the speed of the reaction.

This reaction requires strict control because double bonds are broken which release significant amounts of energy.

A runaway polymer explosion is a concern when monomers are transported in large tanks by truck or by rail. To reduce this concern, chemical inhibitors are mixed into the monomer tank that will block the reaction during shipment. However, if inhibitors are not added in required amounts or if they are released as a result of

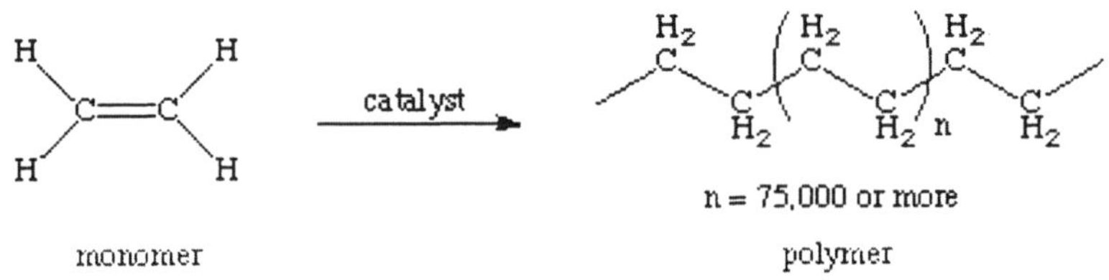

n = 75,000 or more

monomer   polymer

Figure 7.4 Ethylene monomers used to form polyethylene.

emergency conditions, explosion protection will be lost. In this regard, the monomer will react and, in the presence of heat polymerization, will violently progress to form an explosion. Some monomers are easily liquefiable gases and, during transportation or in storage, they are subject to BLEVE.

B = Boiling

L = liquid

E = Expanding

V=Vapor

E = Explosion

Chapter 8

HUMAN HEALTH HAZARDS

PARADIGMS OF HUMAN HEALTH HAZARDS

Today, just about anything that has a potential to negatively affect human health is considered to be a human health hazard. In fact, anything that could potentially have an effect upon the environment may be considered a human health hazard.

Remember the "horror" stories told by your parents and grandparents about breathing dusts in mines and smelters that were loaded with silicon and toxic disease causing materials? Do you remember the Walt Disney fairy tale "Alice in Wonderland"? The Mad Hatter did not have a fictional condition. For instance, people that made felt hats, that colored fabric, painted outer structures, made pewter figurines and filled teeth with metal fillings (or wore the fillings) were exposed to enough lead, mercury, and other substances to become ill with neurological disease symptoms. Yes many developed neurosis or became downright crazy from hazardous materials exposures.

The "baby boomers" or children of the 50's and 60's stood in lines to be inoculated with test vaccines for polio—thousands acquiring the disease from the shot that was supposed to protect them. Almost every neighborhood in the United States had homes with people dying from the crippling affects of polio—it almost became commonplace to have an iron lung pumping air to keep a victimized family member alive. Our mothers were provided with a wonder drug to help reduce pregnancy illness only to learn about a new term called "teratogen" as their babies were born with the most horrible birth defects ever witnessed by the world. Oh yes, the same baby boomers enjoyed sitting on the front porch of our homes while waiting for the musical ice cream truck to stop and we would wave at the nice men that daily sprayed DDT and Chlordane through our neighborhoods to control polio carrying flies and encephalitis carrying mosquitoes. At the pinnacle of our lives, hundreds of thousands of the healthiest American baby boomers were selected by our friends and neighbors to fight communism in the jungles of Southeast Asia, to receive gratitude from the world tobacco producers who provided us with free cigarettes, and to be sprayed with a substance called "Agent Orange."

Of those that remain, many are business owners, scientists, technicians, and leaders in the world. Many of us have set goals to tell our posterity about what may have influenced today's world to be as crazy as it may appear, and to help others to learn from history—so they may not be forced to live it over again.

There are thousands of professionals that work daily in the United States and across the globe to assess, evaluate, and control potential human health hazards. Looking for an environmental or health-related job? In the United States you may try the United States Environmental Protection Agency (EPA), the Occupational Safety and Health Administration (OSHA), Centers for Disease Control (CDC), National Institute of Occupational Safety and Health (NIOSH), the World Health Organization (WHO), public health agencies, state environmental & health agencies, city environmental and health agencies, hospitals, laboratory networks, federal emergency management (FEMA), state and local emergency management and response agencies, law enforcement, and industrial environmental, health, and safety entities within most companies.

In the past, how have we determined that something poses a human health threat?

Unfortunately, politics and special interests have played a part de-emphasizing hazardous considerations and on the other hand other special interests promote uncontrolled over-reactions. The term HAZMAT is a key story selling topic for many news anchors—on all networks. In this chapter, we'll talk about who is involved in characterizing human health hazards and we will look at lists of potentially dangerous substances that the regulatory agencies have shown a bright orange light upon to raise our awareness and to stimulate our consideration. *Truth and knowledge is golden....*

CHARACTERIZATION AND ASSIGNMENT OF HEALTH HAZARD STATUS

Public health hazards characterized by the U.S. Department of Health and Human Services, specifically the Agency for Toxic Substances and Disease Registry (ATSDR), are classified in three prime areas with five categories as listed below:
(Reference—CDC/ATSDR PHAP Classifications 11/30/04)

A. Substances that definitely pose health hazards are categorized as:
   • Category 1: Urgent public health hazard
   • Category 2: Known public health hazard

B.   Substances that cannot be fully evaluated because *critical* information about them is missing or is not yet determined:
   • Category 3: Indeterminate public health hazard

C.   Potential health hazards that are suspect, but at the present time:
   • Category 4: Are not causing apparent public health concerns
   • Category 5: Have not presented a known public health hazard

Category 1—Urgent Public Health Hazards

This category is used when short-term exposures (< 1 yr) to certain hazardous substances or hazardous conditions can present serious health conditions within a given population.
   For example:

• Specific conditions exist, or have a potential to exist, that require immediate attention in order to prevent disastrous effects upon a given population.
• Hazardous conditions exist that involve the presence of serious physical or safety hazards, e.g. open mine shafts, containers, or devices that could release radioactive or toxic substances.

Category 1—Urgent Public Health Hazards may require immediate action by ATSDR and local health agencies for implementing investigations. Methods of investigation may include conducting:

• biologic monitoring
• biomedical testing
• case studies
• epidemiologic studies
• community health investigations
• public health surveillance
• cluster investigations
• health statistics reviews
• health professional education
• community health education
• substance-specific applied research
• reviews of registries

Category 2—Public Health Hazard

Category 2 Public Health Hazards involve substances and related conditions that have a potential to affect a population after having long-term exposures or exposures that exceed one year.

For example:

- Specific health conditions occurring in a population that are the result from long term dangerous substance exposure
- As in Category 1, may also include physical conditions that can affect a given population
- Category 2—Public Health Hazards would also stimulate intervention by federal and local public health agencies

Category 3—Indeterminate Public Health Hazard

This category is used in situations that critical information is lacking to pinpoint actual public health conditions. For example:

- Not enough proof that critical health conditions are potential
- Enough concern warrants opening investigations because of complaints and public health concerns—some limited data may indicate further investigation is required

SAFETY AND HEALTH CONSIDERATIONS

The United States Occupational Safety and Health Administration (OSHA) has done a great job to define and redefine issues relating to controlling human health hazards in the workplace. The author strongly exhorts the readers of this text to routinely review the OSHA and NIOSH/CDC websites to keep up-to-date with guidance information and research results provided by these organizations. Why? They belong to you; your tax dollars are spent daily to keep them working for you and they are extremely valuable to all of us. In this regard, the following human health hazard summaries and definitions are directly quoted from the OSHA Hazard Communication directive, 29 CFR 1910.1200 Appendix E—a prime directive used in characterizing human health hazards.

Although safety hazards related to the physical characteristics of a chemical can be objectively defined in terms of testing requirements (e.g. flammability), health hazard definitions are less precise and more subjective. Health hazards may cause measurable changes in the body—such as decreased pulmonary function. These changes are generally indicated by the occurrence of signs and symptoms in the exposed employees - such as shortness of breath, a non-measurable, subjective feeling. Employees exposed to such hazards must be apprised of both the change in body function and the signs and symptoms that may occur to signal that change.

The determination of occupational health hazards is complicated by the fact that many of the effects or signs and symptoms occur commonly in non-occupationally exposed populations, so that effects of exposure are difficult to separate from normally occurring illnesses. Occasionally, a substance causes an effect that is rarely seen in the population at large, such as angiosarcomas caused by vinyl chloride exposure, thus making it easier to ascertain that the occupational exposure was the primary causative factor. More often, however, the effects are common, such as lung cancer. The situation is further complicated by the fact that most chemicals have not been adequately tested to determine their health hazard potential, and data do not exist to substantiate these effects.

There have been many attempts to categorize effects and to define them in various ways. Generally, the terms "acute" and "chronic" are used to delineate between effects on the basis of severity or duration. "Acute" effects usually occur rapidly as a result of short-term exposures and are of short duration. "Chronic" effects generally occur as a result of long-term exposure and are of long duration.

The acute effects referred to most frequently are those defined by the American National Standards Institute (ANSI) standard for Precautionary Labeling of Hazardous Industrial Chemicals (Z129.1-1988) - irritation, corrosivity, sensitization, and lethal dose. Although these are important health effects, they do not adequately cover the considerable range of acute effects which may occur as a result of occupational exposure, such as, for example, narcosis.

Similarly, the term chronic effect is often used to cover only carcinogenicity, teratogenicity, and mutagenicity. These effects are obviously a concern in the workplace; but again, do not adequately cover the area of chronic effects, excluding, for example, blood dyscrasias (such as anemia), chronic bronchitis, and liver atrophy.

The goal of defining precisely, in measurable terms, every possible health effect that may occur in the workplace as a result of chemical exposures cannot realistically be accomplished. This does not negate the need for employees to be informed of such effects and protected from them. Appendix B, which is also mandatory, outlines the principles and procedures of hazard assessment.

For purposes of this section, any chemicals which meet any of the following definitions, as determined by the criteria set forth in Appendix B, are health hazards. However, this is not intended to be an exclusive categorization scheme. If there is available scientific data that involve other animal species or test methods, they must also be evaluated to determine the applicability of the HCS.

1. "Carcinogen:" A chemical is considered to be a carcinogen if:

   (a) It has been evaluated by the International Agency for Research on Cancer (IARC), and found to be a carcinogen or potential carcinogen; or

   (b) It is listed as a carcinogen or potential carcinogen in the Annual Report on Carcinogens published by the National Toxicology Program (NTP) (latest edition); or,

   (c) It is regulated by OSHA as a carcinogen.

2. "Corrosive:" A chemical that causes visible destruction of, or irreversible alterations in, living tissue by chemical action at the site of contact. For example, a

chemical is considered to be corrosive if, when tested on the intact skin of albino rabbits by the method described by the U.S. Department of Transportation in Appendix A to 49 CFR part 173, it destroys or changes irreversibly the structure of the tissue at the site of contact following an exposure period of four hours. This term shall not refer to action on inanimate surfaces.

3. "Highly toxic:" A chemical falling within any of the following categories:

(a) A chemical that has a median lethal dose (LD(50)) of 50 milligrams or less per kilogram of body weight when administered orally to albino rats weighing between 200 and 300 grams each.

(b) A chemical that has a median lethal dose (LD(50)) of 200 milligrams or less per kilogram of body weight when administered by continuous contact for 24 hours (or less if death occurs within 24 hours) with the bare skin of albino rabbits weighing between two and three kilograms each.

(c) A chemical that has a median lethal concentration (LC(50)) in air of 200 parts per million by volume or less of gas or vapor, or 2 milligrams per liter or less of mist, fume, or dust, when administered by continuous inhalation for one hour (or less if death occurs within one hour) to albino rats weighing between 200 and 300 grams each.

4. "Irritant:" A chemical which is not corrosive, but which causes a reversible inflammatory effect on living tissue by chemical action at the site of contact. A chemical is a skin irritant if, when tested on the intact skin of albino rabbits by the methods of 16 CFR 1500.41 for four hours exposure or by other appropriate techniques, it results in an empirical score of five or more. A chemical is an eye irritant if so determined under the procedure listed in 16 CFR 1500.42 or other appropriate techniques.

5. "Sensitizer:" A chemical that causes a substantial proportion of exposed people or animals to develop an allergic reaction in normal tissue after repeated exposure to the chemical.

6. "Toxic." A chemical falling within any of the following categories:

(a) A chemical that has a median lethal dose (LD(50)) of more than 50 milligrams per kilogram but not more than 500 milligrams per kilogram of body weight when administered orally to albino rats weighing between 200 and 300 grams each.

(b) A chemical that has a median lethal dose (LD(50)) of more than 200 milligrams per kilogram but not more than 1,000 milligrams per kilogram of body weight when administered by continuous contact for 24 hours (or less if death occurs within 24 hours) with the bare skin of albino rabbits weighing between two and three kilograms each.

(c) A chemical that has a median lethal concentration (LC(50)) in air of more than 200 parts per million but not more than 2,000 parts per million by volume of gas or vapor, or more than two milligrams per liter but not more than 20 milligrams per liter of mist, fume, or dust, when administered by continuous inhalation for one hour (or less if death occurs within one hour) to albino rats weighing between 200 and 300 grams each.

7. "Target organ effects."

The following is a target organ categorization of effects which may occur, including examples of signs and symptoms and chemicals which have been found to cause such effects. These examples are presented to illustrate the range and diversity of effects and hazards found in the workplace, and the broad scope employers must consider in this area, but are not intended to be all-inclusive.

a. Hepatotoxins: Chemicals which produce liver damage

Signs & Symptoms: Jaundice; liver enlargement

Chemicals: Carbon tetrachloride; nitrosamines

b. Nephrotoxins: Chemicals which produce kidney damage

Signs & Symptoms: Edema; proteinuria

Chemicals: Halogenated hydrocarbons; uranium

c. Neurotoxins: Chemicals which produce their primary toxic effects on the nervous system

Signs & Symptoms: Narcosis; behavioral changes; decrease in motor functions

Chemicals: Mercury; carbon disulfide

d. Agents which act on the blood or hemato-poietic system: Decrease hemoglobin function; deprive the body tissues of oxygen

Signs & Symptoms: Cyanosis; loss of consciousness

Chemicals: Carbon monoxide; cyanides

e. Agents which damage the lung: Chemicals which irritate or damage pulmonary tissue

Signs & Symptoms: Cough; tightness in chest; shortness of breath

Chemicals: Silica; asbestos

f. Reproductive toxins: Chemicals which affect the reproductive capabilities including chromosomal damage (mutations) and effects on fetuses (teratogenesis)

Signs & Symptoms: Birth defects; sterility

Chemicals: Lead; DBCP

g. Cutaneous hazards: Chemicals which affect the dermal layer of the body

Signs & Symptoms: Defatting of the skin; rashes; irritation

Chemicals: Ketones; chlorinated compounds

h. Eye hazards: Chemicals which affect the eye or visual capacity

Signs & Symptoms: Conjunctivitis; corneal damage

Chemicals: Organic solvents; acids

LINGERING COMMUNITY AND INDUSTRIAL HEALTH CONCERNS

Due to politics, improper business practices, and needs for specific processes, several sources of hazardous substance health hazards require additional control. Two substances of special interest are mercury and chromium.

What About Mercury (Hg)?

Mercury exposures may occur when people come in contact with it blended into certain products—years ago the product called ammoniated mercury was commonly used to treat skin infections. It is found as a mixture in fluorescent lamps and in the high powered mercury vapor lamps used at ball parks around the world. Have you been to see a stage play lately? The bright lamps that light the stage usually contain mercury vapors as well. In its pure form (liquid metal) mercury is used in various kinds of thermometers, electrical switches, contactors, and more. Mercury (liquid) metal emits toxic vapors. In the past, people enjoyed rolling it around their hands or rubbing it onto a coin so the coin would brilliantly shine.

Pathways for Exposures to Mercury

Mercury exposures occur using the standard format:

- Breathing it into the lungs and nasal mucosa (inhalation)
- Eating it or swallowing contaminants collected into the nasal mucosa (ingestion)
- Absorbing it though the skin (unbroken skin) (absorption)
- Open wounds, cracks in the skin or punctures of the skin (injection /skin damage)

By the way, do you like to eat fish? It is a brain food—please know that today we are concerned about a product often found in fish that also affects the brain, methylmercury. Methyl mercury (an organic complex that is easy to absorb into the human body) is more toxic than mercury metal—it collects into the tissues and oils of fish. How does it get there? Remember, the Environmental Protection Agency is an American agency—most of our world neighbors do not share our environmental protection interests. What about the United Nations? We better keep checking the fish.

Mercury Effects Upon the Human Body

The presence of mercury can cause diseases of the:

- Human brain
- Spinal cord
- Kidneys
- Lungs
- Liver

Exposure symptoms can include nausea, shortness of breath, fever, muscle aches, sore gums, and increased amounts white blood cells:

- Short-term exposure symptoms may include tingling or numbness of your fingers and toes. Some people notice numbing around the mouth, and the shakes or tremors. If not properly tested, a physician could mistake the symptoms for the flu or diabetes.
- Long-term exposure symptoms to significant levels of mercury may produce progressively worse symptoms, some people undergo changes in their personality, tunnel vision, incoherence, neurogenic disorders, "vegetative states of being," and eventual coma.
- Also worth noting, mercury can have an effect upon unborn children by preventing the brain and nervous system from properly developing. Children that have been exposed to mercury show decreased intellectual activity, hearing impairment, and impaired coordination.

What about Chromium (Cr) Exposure?

Chrome-containing compounds, especially +6 valences or the hexavalent form, is considered to be a significant human health hazard. People that perform activities in metal finishing, electroplating, spray painting, surface coating, and welding, are subject to extremes of chromium exposure.

What can chromium do to you? It's also considered to be a mineral that is helpful if taken in trace amounts in our diet! Not hexavalent chrome—that stuff can be bad news.

Chrome Health Effects

According to OSHA, NIOSH, and CDC, calcium chromate, chromium trioxide, lead chromate, strontium chromate, and zinc chromate have been determined to cause cancer in humans. Lung cancer appears to be very prevalent among workers that produce chromate and manufacture pigments containing chromate. Also lung cancer appears prevalent in consumers of chromate pigments. Foundry workers that work with chromium-nickel alloys increase in lung cancers. People that work with chromic acid and associated chrome-containing products are at risk. The following Human Health Hazard information is provided by OSHA:

Chrome Hazard Summary

1. Known Source of Cancer:
   - Hexavalent chromium is considered a potential lung carcinogen. Studies of workers in the chromate production, plating, and pigment industries consistently show increased rates of lung cancer.
2. Permanent Eye Damage:
   - Direct eye contact with chromic acid or chromate dusts can cause permanent eye damage.

3. Respiratory Tract Damage:

- Hexavalent chromium can irritate the nose, throat, and lungs. Repeated or prolonged exposure can damage the mucous membranes of the nasal passages and result in ulcers. In severe cases, exposure causes perforation of the septum (the wall separating the nasal passages). Acute exposures may cause perforation of the nasal septum within a week of exposure.
- *Occupational Dermatoses*. Refer to work reported by the National Institute for Occupational Safety and Health (NIOSH) (2001, April 17).

4. Skin Damage:

- Prolonged skin contact can result in dermatitis and skin ulcers. Some workers develop an allergic sensitization to chromium. In sensitized workers, contact with even small amounts can cause a serious skin rash. Kidney damage has been linked to high dermal exposures.
- *Occupational Dermatoses*. National Institute for Occupational Safety and Health (NIOSH) (2001, April 17).

ENVIRONMENTAL HEALTH CONCERNS—PROTECT THE CHILDREN

A prime responsibility that the U.S. Environmental Protection Agency has to deal with is protecting Americans as well as the world community from environmental risks. A major concern for EPA has been to develop programs that would focus upon protecting children from environmental hazards. An agency-wide policy was developed in 1995 to ensure that environmental health risks of children are explicitly and consistently evaluated and coordinated with public health standards. In 1996 a National Agenda to Protect Children's Health From Environmental Threats expanded the Agency's activities. It is aimed at ensuring a consistent approach to improving risk assessments and to specifically address children's risks. EPA committed to establish programs that will protect all children from environmental health threats and the agency is working to further develop national health-based environmental standards. The following information has been provided by the U.S. Environmental Protection Agency for our review (Reference: EPA Region 8, Environmental Health Hazards 7/2004—4/5/2005):

Asbestos-Induced Cancer

Asbestos is a fibrous, naturally-occurring mineral that over the years was used to strengthen many products, was used as an insulator, and to protect materials and equipment from fire. Asbestos has been used in many construction materials, such as in products used for roofing, siding, plumbing and heater insulation, floor tiles, ceiling panels, coatings, and gaskets. It is often found in structures built before the mid 1980's and in some situations it is currently being used. Asbestos fibers easily penetrate tissues in most parts of the respiratory system. When it becomes lodged within the lungs, it is difficult to expel. It can remain within the lungs for long periods of

Table 8.1   Properties of Hexavalent Chromium Compounds.

	Calcium Chromate	Chromium Trioxide	Lead Chromate	Potassium Dichromate	Strontium Chromate	Zinc Chromate
Physical Description	yellow monoclinic prisms	odorless, dark-purplish to red rhombus crystals that are deliquescent; when heated to decomposition, chromium trioxide emits smoke and irritating fumes	yellow or orange monoclinic crystals; when heated to decomposition, emit toxic fumes of lead; basic lead chromate is a red amorphous or crystalline powder	orange-red crystals	monoclinic yellow crystals	lemon yellow prisms
Molecular Weight (grams/mol)	156.01	99.99	323.18	294.21	203.62	181.97
Melting Point	No Data	196°C	844°C	398°C	No Data	No Data
Vapor Pressure	N/A (solid form)	N/A (solid form)	N/A (solid form)	N/A (solid form)	N/A (solid form)	N/A (solid form)
Specific Gravity	2.89 g/cm_	2.70 g/cm_	6.30 g/cm_	2.68 g/cm_	3.89 g/cm_	3.40 g/cm_
Solubility	soluble in cold and hot water and reacts with acids and ethanol	soluble in alcohol, ethanol, sulfuric acid, and nitric acid	insoluble in water, acetic acid, and ammonia; soluble in acid and alkali	soluble in water, acid, and alkali	soluble in water; reacts with hydrochloric acid, nitric acid, acetic acid, and ammonium salts	insoluble in cold water and acetone; dissolves in hot water; soluble in acid and liquid ammonia
Flammability	N/A	N/A	N/A	N/A	N/A	N/A

Occupational Safety and Health Administration, Revised: 13 January 2004

time, leading to increased potential for lung and respiratory system cancer. Asbestos-contaminated people that are routinely exposed to chemical fumes or that may smoke tobacco products suffer radically increased risks for respiratory system cancers.

Asthma Concerns

Asthma is a chronic respiratory disease that causes narrowing airways within the respiratory system. Symptoms include coughing, wheezing, chest tightness, shortness of breath, and increased risk of respiratory infections. Asthma attacks can be "triggered" by irritants and allergens such as cigarette smoke, wood smoke, smog, pollen, household dust mites, pet dander, cockroaches, and molds. Most children and adults with asthma have significant allergies to these environmental triggers. Asthma is a leading cause of long-term illness in children. Please note that both the U.S EPA and OSHA website contains more valuable information about controlling indoor air quality.

Toxic Chemicals, Biocides, and Pesticides

Many toxic chemicals are used within industry and in household cleaning products. Adults may withstand exposures to some chemicals where children may quickly develop serious health conditions. It should also be mentioned that teratogenic chemicals, often used in cleaners and paints, have the potential to cross the placenta in an expectant mother and can cause serious birth defects in an unborn child.

Biocides and pesticides are substances intended to destroy, control, or repel pests, such as insects, weeds, rodents, and microbiological contaminants. These kinds of chemicals may cause diseases such as cancer, lung disorders, nervous, reproductive, endocrine, and immune system damage and some may accumulate in the environment. Children are at greater risk of exposure than adults because, pound for pound of body weight, children not only eat more and breathe more, but they also have a more rapid metabolism than adults. Children play on floors and outdoors where biocides or pesticides are used.

Recommendations to prevent exposures by children:

- Store food and trash in closed containers to keep pests from coming indoors
- Refrain from using pesticides if possible—seek safe alternatives
- Read product labels, MSDS, and follow manufacturer's directions
- Use bait & traps for insects and rodents
- Store chemicals in locations where children cannot reach them
- Keep biocides, pesticides, and herbicides away from children—don't allow children to enter areas that have been recently chemically treated
- Wash produce with copious amounts of water; peel whenever possible

Consult the EPA website for more information about preventing children's exposures to toxic chemicals and pesticides.

Radon Gas Hazards

Radon gas is a radioactive gas that is naturally emitted from below the earth's surface and travels through rock and soil formations as it is released into the atmosphere. It is odorless, tasteless, and colorless. Breathing air containing radon carries the risk of developing lung cancer. The amount of risk depends on the radon concentration in the air and total time of exposure (years). Next to smoking, radon may be considered a leading cause of lung cancer deaths each year. EPA recommends that indoor air concentrations of radon should be kept less than 4 picoCuries per Liter (pCi/L) of air. EPA also estimates about 10 percent of homes and buildings in the United States have indoor air concentrations of radon well above the danger limit. Radon test kits are available from many resources; most hardware stores sell them at a reasonable price to test your home or business. Consult the U.S. EPA website for more information on radon.

Tobacco Smoke

According to the U.S. EPA, tobacco smoke contains more that 4,000 chemicals, which include carbon monoxide, nicotine, tars, formaldehyde, and hydrogen cyanide. People that breathe secondhand smoke are forced to involuntarily breathe chemical pollutants. Several of these substances cause heart damage, respiratory system damage and cancer. Children exposed to secondhand smoke are prone to bronchitis, pneumonia, respiratory infections, ear infections, and asthma symptoms. Frequencies of respiratory infection can be directly related to the amount of smoke in the home. As one may expect, this issue has several political overtones. If smoking is allowed within the workplace, please know that laws will eventually become enforced against allowing this practice. Business owners should also know that by allowing tobacco smoking in the workplace, this practice may eventually be construed as purposely exposing workers to known hazards without protection. A good lawyer may be able to address this issue better. As far as exposing children to tobacco smoke—don't smoke around them.

Protect Children from Lead

OSHA tightly regulates potential lead exposure in the workplace. EPA believes that lead is a major environmental health hazard for young children. Approximately 75 percent of homes and buildings built in the United State before 1978 contain lead paint and other lead contamination sources. Lead demonstrates similar hazards as mercury and they can cause neurological disorders in all people. Children that live in older homes are exposed to increased lead exposure potentials from chipping or peeling lead paint; and by lead-contaminated dusts from paint removed during remodeling. Lead acetate in some paints taste sweet to some children and they will often eat chipped lead paint. Drinking water from lead-soldered plumbing may affect children. Children that suffer from lead exposure may demonstrate symptoms of blood and kidney disorders, neurological damage, and learning disorders.

Hazards from Ultraviolet (UV) Radiation

Overexposure to the sun's harmful UV light may damage children's skin, cause eye damage, and suppress the immune system. UV light is the sun's radiation which has increased on the earth's surface due to damage to the earth's ozone layer in the upper atmosphere. Children 10-15 years of age who have experienced excessive sunburns are three times as likely to develop malignant melanoma, the most deadly kind of skin cancer, later in life.

EPA Recommendation: Keep children out of the midday sun; always use sunscreen (with a sun protection factor [SPF] of at least 15); and wear light clothing, hats, and sunglasses.

Water Contamination

Water-damaged materials and pools of stagnant water that have remained for longer than 48 hours should be considered contaminated sources for illness. Sewer backup, flood waters and the like are considered to hold the highest potential for pathogenic (disease-causing) microbial contamination. It is commonly known that water towers and places that are kept wet all the time are potential sources for legionnaire's disease that is caused by the bacteria legionnella. Don't forget mold and other potentially hazardous water borne fungal infections that also result from water damage, post fire clean up and other conditions. EPA is also concerned about the quality of water that adults and children may drink or be exposed to.

Children's exposure to waterborne contaminants can occur from drinking water, eating contaminated fish, or swallowing water while swimming in contaminated lakes, oceans, or streams:

- Drinking water: Nitrates are a result of the use of chemical fertilizers which then leach into the water. Consuming small amounts of nitrate is not harmful, but larger amounts are toxic to infants and cause oxygen depletion in the body.
- Fish consumption: Pollutants, such as mercury that accumulates in the tissues of fish, are of great concern because of the potential to cause birth defects, liver damage, cancer, and other serious health problems.
- Recreation: Disease-causing organisms in sewage-contaminated water can cause hepatitis, dysentery, fever, gastrointestinal illness, ear infections, and other health problems.

EPA recommendations to prevent exposure to waterborne hazards:

- Drinking Water: If you suspect your drinking water has questionable nitrate concentrations, contact your local water supplier and ask for the test results or have your private well tested.

- Fish Consumption: Because unborn children may be most sensitive to the effects of mercury, women of child-bearing age should limit their consumption of marine and freshwater fish taken from contaminated waters.
- Recreation: Don't allow children to swim near areas of questionable water quality. For more information in this regard, visit the EPA website for the region that you reside.

Table 8.2 Listed chemical health hazards. *(continued on next page)*

Chemical Name	CAS Registry Number (or EDF Substance ID)
ACETALDEHYDE	75-07-0
ACETAMIDE	60-35-5
ACETAZOLAMIDE	59-66-5
ACETOCHLOR	34256-82-1
ACETOHYDROXAMIC ACID	546-88-3
2-ACETYLAMINOFLUORENE	53-96-3
ACETYLSALICYLIC ACID	50-78-2
ACIFLUORFEN, SODIUM SALT	62476-59-9
ACRYLAMIDE	79-06-1
ACRYLONITRILE	107-13-1
ACTINOMYCIN D	50-76-0
ADRIAMYCIN	23214-92-8
AF-2; (2-[2-FURYL]-3-[5-NITRO-2-FURYL])ACRYLAMIDE	3688-53-7
AFLATOXINS	1402-68-2
AIREDALE BLUE D	2429-74-5
ALACHLOR	15972-60-8
ALCOHOLIC BEVERAGES, WHEN ASSOCIATED WITH ALCOHOL ABUSE	EDF-008
ALDRIN	309-00-2
ALL-TRANS RETINOIC ACID	302-79-4
1-ALLYL-4-METHOXYBENZENE	140-67-0
ALPHA-NAPHTHYLAMINE	134-32-7
ALPRAZOLAM	28981-97-7
AMANTADINE HYDROCHLORIDE	665-66-7
AMIKACIN SULFATE	39831-55-5
1-AMINO-2,4-DIBROMOANTHRAQUINONE	81-49-2
1-AMINO-2-METHYLANTHRAQUINONE	82-28-0
4-AMINO-2-NITROPHENOL	119-34-6
2-AMINO-3-METHYL-9H-PYRIDO (2,3-B) INDOLE	68006-83-7
2-AMINO-5-(5-NITRO-2-FURYL)-1,3,4-THIADIAZOLE	712-68-5
2-AMINO-6-METHYLDIPYRIDO (1,2-A:3',2'-D) IMIDAZOLE	67730-11-4
3-AMINO-9-ETHYLCARBAZOLE HYDROCHLORIDE	6109-97-3
2-AMINO-9H-PYRIDO (2,3-B) INDOLE (A-ALPHA-C)	26148-68-5
2-AMINO-DIPYRIDO (1,2-A:3',2'-D)-IMIDAZOLE	67730-10-3
2-AMINOANTHRAQUINONE	117-79-3
4-AMINOAZOBENZENE	60-09-3
4-AMINOBIPHENYL	92-67-1
2-AMINOFLUORENE	153-78-6
AMINOGLUTETHIMIDE	125-84-8
AMINOGLYCOSIDES	EDF-076
2-AMINONAPHTHALENE	91-59-8

Table 8.2 Listed chemical health hazards. *(continued from previous page)*

Chemical Name	CAS Registry Number (or EDF Substance ID)
AMINOPTERIN	54-62-6
AMIODARONE HYDROCHLORIDE	19774-82-4
AMITRAZ	33089-61-1
AMITROLE	61-82-5
AMOXAPINE	14028-44-5
ANABOLIC STEROIDS	EDF-006
ANALGESIC MIXTURES CONTAINING PHENACETIN	EDF-011
ANGIOTENSIN CONVERTING ENZYME (ACE) INHIBITORS	1407-47-2
ANILINE	62-53-3
ANILINE HYDROCHLORIDE	142-04-1
ANISINDIONE	117-37-3
ANTIMONY TRIOXIDE	1309-64-4
ARAMITE	140-57-8
ARISTOLOCHIC ACID	313-67-7
ASBESTOS (FRIABLE)	1332-21-4
ASPHALT (PETROLEUM) FUMES	8052-42-4
ATENOLOL	29122-68-7
ATTAPULGITE	12174-11-7
AURAMINE	492-80-8
AURANOFIN	34031-32-8
5-AZACYTIDINE	320-67-2
AZASERINE	115-02-6
AZATHIOPRINE	446-86-6
AZOBENZENE	103-33-3
BARBITURATES	BAG500
BECLOMETHASONE DIPROPIONATE	5534-09-8
BENOMYL	17804-35-2
BENZ(A)ANTHRACENE	56-55-3
BENZENE	71-43-2
BENZIDINE	92-87-5
BENZO(A)PYRENE	50-32-8
BENZO(B)FLUORANTHENE	205-99-2
BENZO (K) FLUORANTHENE	207-08-9
BENZODIAZEPINES	EDF-007
BENZOFURAN	271-89-6
BENZOIC TRICHLORIDE	98-07-7
BENZO (J) FLUORANTHENE	205-82-3
BENZPHETAMINE HYDROCHLORIDE	5411-22-3
BENZYL CHLORIDE	100-44-7
BENZYL VIOLET 4B	1694-09-3
BERYLLIUM	7440-41-7

Table 8.2 Listed chemical health hazards. *(continued from previous page)*

Chemical Name	CAS Registry Number (or EDF Substance ID)
BERYLLIUM COMPOUNDS	BFQ500
BETA-BUTYROLACTONE	3068-88-0
BETA-PROPIOLACTONE	57-57-8
BETEL QUID WITH TOBACCO	EDF-013
1,1'-BI (ETHYLENE OXIDE)	1464-53-5
4',4'''-BIACETANILIDE	613-35-4
(1,1'-BIPHENYL)-4,4'-DIAMINE, 3,3'-DIMETHYL-	119-93-7
(1,1'-BIPHENYL)-4,4'-DIAMINE, 3,3'-DIMETHYL-, DIHYDROCHLORIDE (9CI)	612-82-8
BIS (2-CHLORO-1-METHYLETHYL) ETHER	108-60-1
BIS (2-CHLOROETHYL) ETHER	111-44-4
BIS (2-ETHYLHEXYL) PHTHALATE	117-81-7
BIS (CHLOROMETHYL) ETHER	542-88-1
BISCHLOROETHYL NITROSOUREA (BCNU)	154-93-8
BRACKEN FERN	BML000
BROMACIL LITHIUM SALT (2,4 (H,3H]-PYRIMIDINEDIONE, ETHYL-3 (1-METHYLPROPYL], LITHIUM SALT)	53404-19-6
BROMATE	15541-45-4
BROMOXYNIL	1689-84-5
BROMOXYNIL OCTANOATE	1689-99-2
BUTABARBITAL SODIUM	143-81-7
1,3-BUTADIENE	106-99-0
1,4-BUTANEDIOL DIMETHANESULFONATE (MYLERAN)	55-98-1
BUTYLATED HYDROXYANISOLE (BHA)	25013-16-5
C.I. ACID RED 114	6459-94-5
C.I. BASIC RED 9 MONOHYDROCHLORIDE	569-61-9
C.I. DIRECT BLUE 218	28407-37-6
C.I. DIRECT BROWN 95	16071-86-6
C.I. FOOD RED 15	81-88-9
C.I. FOOD RED 5	3761-53-3
C.I. SOLVENT YELLOW 14	842-07-9
C.I. SOLVENT YELLOW 3	97-56-3
CACODYLIC ACID	75-60-5
CADMIUM	7440-43-9
CADMIUM COMPOUNDS	CAE750
CAFFEIC ACID	331-39-5
CAMPHECHLOR	8001-35-2
CAPTAFOL	2425-06-1
CAPTAN	133-06-2
CARBAMAZEPINE	298-46-4
CARBAZOLE	86-74-8

Table 8.2   Listed chemical health hazards. *(continued from previous page)*

Chemical Name	CAS Registry Number (or EDF Substance ID)
CARBON BLACK	1333-86-4
CARBON DISULFIDE	75-15-0
CARBON MONOXIDE	630-08-0
CARBON TETRACHLORIDE	56-23-5
CARBOPLATIN	41575-94-4
CARBOXYMETHYLNITROSOUREA	60391-92-6
CATECHOL	120-80-9
CERAMIC FIBERS (AIRBORNE PARTICLES OF RESPIRABLE SIZE)	1056
CERTAIN COMBINED CHEMOTHERAPY FOR LYMPHOMAS	EDF-016
CHENODIOL	474-25-9
CHINOMETHIONAT (6-METHYL-1,3-DITHIOLO(4,5-B] QUINOX	2439-01-2
2-CHLOR-1,3-BUTADIENE	126-99-8
CHLORAMBUCIL	305-03-3
CHLORAMPHENICOL	56-75-7
CHLORCYCLIZINE HYDROCHLORIDE	1620-21-9
CHLORDANE	57-74-9
CHLORDECONE (KEPONE)	143-50-0
CHLORDIAZEPOXIDE	58-25-3
CHLORDIAZEPOXIDE HYDROCHLORIDE	438-41-5
CHLORDIMEFORM	6164-98-3
CHLORENDIC ACID	115-28-6
CHLORINATED PARAFINS (AVERAGE CHAIN LENGTH, C12; APPROXIMATELY 60 PERCENT CHLORINE BY WEIGHT)	108171-26-2
3-CHLORO-2-METHYL-1-PROPENE	563-47-3
5-CHLORO-O-TOLUIDINE AND ITS STRONG ACID SALTS	EDF-216
4-CHLORO-ORTHO-PHENYLENEDIAMINE	95-83-0
CHLOROBENZILATE	510-15-6
CHLOROETHANE	75-00-3
1-(2-CHLOROETHYL)-3-(4-METHYLCYCLOHEXYL)-1-NITROSOUREA (METHYL CCNU)	13909-09-6
1-(2-CHLOROETHYL)-3-CYCLOHEXYL-1-NITROSOUREA	13010-47-4
CHLOROFORM	67-66-3
CHLOROMETHANE	74-87-3
CHLOROMETHYL METHYL ETHER	107-30-2
CHLOROTHALONIL	1897-45-6
CHLOROTRIANISENE	569-57-3
CHLOROZOTOCIN	54749-90-5
CHLORSULFURON	64902-72-3

Table 8.2   Listed chemical health hazards. *(continued from previous page)*

Chemical Name	CAS Registry Number (or EDF Substance ID)
CHROMIUM (VI) COMPOUNDS	18540-29-9
CHRYSENE	218-01-9
CICLOSPORIN (CYCLOSPORIN A; CYCLOSPORINE)	79217-60-0
CIDOFOVIR	113852-37-2
CINNAMYL ANTHRANILATE	87-29-6
CISPLATIN	15663-27-1
CITRUS RED NO.2	6358-53-8
CLADRIBINE	4291-63-8
CLARITHROMYCIN	81103-11-9
CLOBETASOL PROPIONATE	25122-46-7
CLOFIBRATE	637-07-0
CLOMIPHENE CITRATE	50-41-9
CLORAZEPATE DIPOTASSIUM	57109-90-7
COBALT	7440-48-4
COBALT SULFATE HEPTAHYDRATE	10026-24-1
COBALT (II) OXIDE	1307-96-6
COCAINE	50-36-2
CODEINE PHOSPHATE	52-28-8
COKE OVEN EMISSIONS	1066
COLCHICINE	64-86-8
CONJUGATED ESTROGENS	EDF-023
CREOSOTES	8001-58-9
CUPFERRON	135-20-6
CYANAZINE	21725-46-2
CYCASIN	14901-08-7
CYCLOATE	1134-23-2
CYCLOHEXIMIDE	66-81-9
CYCLOPHOSPHAMIDE	50-18-0
CYCLOPHOSPHAMIDE (HYDRATED)	6055-19-2
CYTARABINE	147-94-4
CYTEMBENA	21739-91-3
D & C ORANGE NO. 17	3468-63-1
D & C RED NO. 8	2092-56-0
D & C RED NO. 9	5160-02-1
DACARBAZINE	4342-03-4
DAMINOZIDE	1596-84-5
DANAZOL	17230-88-5
DANTRON (CHRYSAZIN; 1,8-DIHYDROXYANTHRAQUINONE)	117-10-2
DAUNOMYCIN	20830-81-3

Table 8.2 Listed chemical health hazards. *(continued from previous page)*

Chemical Name	CAS Registry Number (or EDF Substance ID)
DAUNORUBICIN HYDROCHLORIDE	23541-50-6
2,4-DB	94-82-6
DDD	72-54-8
DDE	72-55-9
DDT	50-29-3
DEMECLOCYCLINE HYDROCHLORIDE (INTERNAL USE)	64-73-3
DI-N-PROPYLNITROSAMINE	621-64-7
2,4-DIAMINOANISOLE	615-05-4
2,4-DIAMINOANISOLE SULFATE	39156-41-7
4,4'-DIAMINODIPHENYL ETHER	101-80-4
4,4'-DIAMINODIPHENYL SULFIDE	139-65-1
2,4-DIAMINOTOLUENE	95-80-7
DIAMINOTOLUENE (MIXED ISOMERS)	25376-45-8
DIAZEPAM	439-14-5
DIAZOXIDE	364-98-7
DIBENZ (A,H) ANTHRACENE	53-70-3
DIBENZOFURANS (CHLORINATED)	1080
DIBENZO (A,E) PYRENE	192-65-4
DIBENZO (A,H) PYRENE	189-64-0
DIBENZO (A,I) PYRENE	189-55-9
DIBENZO (A,L) PYRENE	191-30-0
DIBENZ (A,H) ACRIDINE	226-36-8
DIBENZ (A,J) ACRIDINE	224-42-0
2,3-DIBROMO-1-PROPANOL	96-13-9
1,3-DIBROMO-2,2-DIMETHYLOLPROPANE	3296-90-0
1,2-DIBROMO-3-CHLOROPROPANE (DBCP)	96-12-8
1,2-DIBROMOETHANE	106-93-4
3,3'-DICHLOR-4,4'-DIAMINO-DIPHENYLAETHER	28434-86-8
1,4-DICHLORO-2-BUTENE	764-41-0
2,2-DICHLOROACETIC ACID	79-43-6
1,4-DICHLOROBENZENE	106-46-7
3,3'-DICHLOROBENZIDINE	91-94-1
3,3'-DICHLOROBENZIDINE DIHYDROCHLORIDE	612-83-9
DICHLOROBROMOMETHANE	75-27-4
1,1-DICHLOROETHANE	75-34-3
1,2-DICHLOROETHANE	107-06-2
DICHLOROMETHANE	75-09-2
DICHLOROPHENE	97-23-4
1,2-DICHLOROPROPANE	78-87-5
1,3-DICHLOROPROPENE (MIXED ISOMERS)	542-75-6

Table 8.2   Listed chemical health hazards. *(continued from previous page)*

Chemical Name	CAS Registry Number (or EDF Substance ID)
DICHLORPHENAMIDE	120-97-8
DICHLORVOS	62-73-7
DICLOFOP METHYL	51338-27-3
DICUMAROL	66-76-2
DIELDRIN	60-57-1
DIENESTROL	84-17-3
DIESEL EMISSIONS	9902
DIETHYL SULFATE	64-67-5
1,2-DIETHYLHYDRAZINE	1615-80-1
DIETHYLSTILBESTROL	56-53-1
DIFLUNISAL	22494-42-4
DIGLYCIDYL RESORCINOL ETHER (DGRE)	101-90-6
DIHYDROERGOTAMINE MESYLATE	6190-39-2
DIHYDROSAFROLE	94-58-6
DIISOPROPYL SULFATE	2973-10-6
DILTIAZEM HYDROCHLORIDE	33286-22-5
3,3'-DIMETHOXYBENZIDINE	119-90-4
3,3'-DIMETHOXYBENZIDINE DIHYDROCHLORIDE	20325-40-0
1,1-DIMETHYL HYDRAZINE	57-14-7
DIMETHYL SULFATE	77-78-1
2,2-DIMETHYL-4,4-METHYLENEDIANILINE	838-88-0
4-DIMETHYLAMINOAZOBENZENE	60-11-7
7,12-DIMETHYLBENZ(A)ANTHRACENE	57-97-6
DIMETHYLCARBAMOYL CHLORIDE	79-44-7
1,2-DIMETHYLHYDRAZINE	540-73-8
DIMETHYLVINYLCHLORIDE	513-37-1
DINITROBUTYL PHENOL	88-85-7
3,9-DINITROFLUORANTHENE	22506-53-2
3,7-DINITROFLUORANTHENE	105735-71-5
1,6-DINITROPYRENE	42397-64-8
1,8-DINITROPYRENE	42397-65-9
2,6-DINITROTOLUENE	606-20-2
2,4-DINITROTOLUENE	121-14-2
DINITROTOLUENE (MIXED ISOMERS)	25321-14-6
DINOCAP	39300-45-3
1,4-DIOXANE	123-91-1
DIPHENYLHYDANTOIN (PHENYTOIN), SODIUM SALT	630-93-3
1,2-DIPHENYLHYDRAZINE	122-66-7
DIPROPYL ISOCINCHOMERONATE	136-45-8
DIRECT BLACK 38	1937-37-7

Table 8.2   Listed chemical health hazards. *(continued from previous page)*

Chemical Name	CAS Registry Number (or EDF Substance ID)
DIRECT BLUE 6	2602-46-2
DISODIUM CYANODITHIOIMIDOCARBONATE	138-93-2
DISPERSE BLUE 1	2475-45-8
DIURON	330-54-1
DOXYCYCLINE	564-25-0
DOXYCYCLINE CALCIUM (INTERNAL USE)	94088-85-4
DOXYCYCLINE HYCLATE (INTERNAL USE)	24390-14-5
DOXYCYCLINE MONOHYDRATE (INTERNAL USE)	17086-28-1
ENDRIN	72-20-8
ENVIRONMENTAL TOBACCO SMOKE	1090
EPICHLOROHYDRIN	106-89-8
2,3-EPOXY 1-PROPANOL	556-52-5
1-EPOXYETHYL-3,4-EPIXYCYCLOHEXANE	106-87-6
ERGOTAMINE TARTRATE	379-79-3
ESTRADIOL-17B	50-28-2
ESTRONE	53-16-7
ESTROPIPATE	7280-37-7
ETHENE, FLUORO-	75-02-5
ETHINYLESTRADIOL	57-63-6
ETHIONAMIDE	536-33-4
ETHOPROP	13194-48-4
ETHYL ACRYLATE	140-88-5
ETHYL ALCOHOL IN ALCOHOLIC BEVERAGES	EDF-026
ETHYL BROMIDE	74-96-4
ETHYL DIPROPYLTHIOCARBAMATE	759-94-4
ETHYL METHANESULFONATE	62-50-0
ETHYLBENZENE	100-41-4
ETHYLENE GLYCOL MONOETHYL ETHER	110-80-5
ETHYLENE GLYCOL MONOETHYL ETHER ACETATE	111-15-9
ETHYLENE GLYCOL MONOMETHYL ETHER	109-86-4
ETHYLENE GLYCOL MONOMETHYL ETHER ACETATE	110-49-6
ETHYLENE OXIDE	75-21-8
ETHYLENE THIOUREA	96-45-7
ETHYLENEIMINE	151-56-4
ETODOLAC	41340-25-4
ETOPOSIDE	33419-42-0
ETRETINATE	54350-48-0
FENOXAPROP ETHYL (2-4-6-CHLORO-2-BENZOXAZOLYEN[OXY]PENOXY)PROPANIC ACID, ETHYL ESTER)	66441-23-4

Table 8.2 Listed chemical health hazards. *(continued from previous page)*

Chemical Name	CAS Registry Number (or EDF Substance ID)
FENOXYCARB	72490-01-8
FILGRASTIM	121181-53-1
FLUAZIFOP-BUTYL	69806-50-4
FLUNISOLIDE	3385-03-3
FLUOROACETIC ACID, SODIUM SALT	62-74-8
FLUOROURACIL	51-21-8
FLUOXYMESTRONE	76-43-7
FLURAZEPAM HYDROCHLORIDE	1172-18-5
FLURBIPROFEN	5104-49-4
FLUTAMIDE	13311-84-7
FLUTICASONE PROPIONATE	80474-14-2
FLUVALINATE	69409-94-5
FOLPET	133-07-3
FORMALDEHYDE	50-00-0
FORMYLHYDRAZINO-4-(5-NITRO-2-FURYL) THIAZOLE	3570-75-0
FUMONISIN B1	116355-83-0
FURAN	110-00-9
FURAZOLIDONE	67-45-8
FURMECYCLOX	60568-05-0
FUSARIN C	79748-81-5
GAMMA-LINDANE	58-89-9
GANCICLOVIR SODIUM	82410-32-0
GASOLINE ENGINE EXHAUST	9911
GEMFIBORZIL	25812-30-0
GLASSWOOL FIBERS (AIRBORNE PARTICLES OF RESPIRABLE SIZE)	EDF-029
GLYCIDALDEHYDE	765-34-4
GOSERELIN ACETATE	65807-02-5
GRISEOFULVIN	126-07-8
GYROMITRIN (ACETALDEHYDE METHYLFORMYL HYDRAZONE)	16568-02-8
7H-DIBENZO (C,G) CARBAZOLE	194-59-2
HALAZEPAM	23092-17-3
HALOBETASOL PROPIONATE	66852-54-8
HALOPERIDOL	52-86-8
HALOTHANE	151-67-7
HC BLUE 1	2784-94-3
HEPTACHLOR	76-44-8
HEPTACHLOR EPOXIDE	1024-57-3
HEXACHLOROBENZENE	118-74-1

Table 8.2   Listed chemical health hazards. *(continued from previous page)*

Chemical Name	CAS Registry Number (or EDF Substance ID)
1,2,3,4,5,6-HEXACHLOROCYCLOHEXANE (MIXTURE OF ISOMERS)	608-73-1
HEXACHLORODIBENZODIOXIN	34465-46-8
HEXACHLOROETHANE	67-72-1
HEXAMETHYLPHOSPHORAMIDE	680-31-9
HISTRELIN ACETATE	EDF-344
HYDRAMETHYLNON	67485-29-4
HYDRAZINE	302-01-2
HYDRAZINE SULFATE	10034-93-2
HYDROXYUREA	127-07-1
IDARUBICIN HYDROCHLORIDE	57852-57-0
IFOSFAMIDE	3778-73-2
INDENO (1,2,3-CD) PYRENE	193-39-5
INDIUM PHOSPHIDE	22398-80-7
INORGANIC ARSENIC COMPOUNDS	EDF-222
IODINE-131	10043-66-0
IPRODIONE	36734-19-7
IQ(2-AMINO-3-METHYLIMIDAZO[4,5-F]QUINOLINE)	76180-96-6
IRON DEXTRAN	9004-66-4
ISOBUTYL NITRITE	542-56-3
ISOSAFROLE	120-58-1
ISOTRETINOIN	4759-48-2
ISOXAFLUTOLE	EDF-309
LACTOFEN	77501-63-4
LASIOCARPINE	303-34-4
LEAD	7439-92-1
LEAD ACETATE	301-04-2
LEAD COMPOUNDS	LCT000
LEAD PHOSPHATE	7446-27-7
LEAD SUBACETATE	1335-32-6
LEUPROLIDE ACETATE	74381-53-6
LEVODOPA	59-92-7
LEVONORGESTREL IMPLANTS	797-63-7
LINURON	330-55-2
LITHIUM CARBONATE	554-13-2
LITHIUM CITRATE	919-16-4
LORAZEPAM	846-49-1
LOVASTATIN	75330-75-5
LYNESTRENOL	52-76-6
M-DINITROBENZENE	99-65-0

Table 8.2 Listed chemical health hazards. *(continued from previous page)*

Chemical Name	CAS Registry Number (or EDF Substance ID)
MANCOZEB	8018-01-7
MANEB	12427-38-2
MEBENDAZOLE	31431-39-7
MECHLORETHAMINE	51-75-2
MEDROXYPROGESTERONE ACETATE	71-58-9
MEGESTROL ACETATE	595-33-5
MEIQ (2-AMINO-3,4-DIMETHYLIMIDAZO[4,5-F] QUINOLINE)	77094-11-2
MEIQX(2-AMINO-3,8-DIMETHYLIMIDAZO[4,5-F] QUINOXALI E)	77500-04-0
MELAMINE, HEXAMETHYL-	645-05-6
MELPHALAN	148-82-3
MENOTROPINS	9002-68-0
MEPROBAMATE	57-53-4
MERCAPTOPURINE	6112-76-1
MERCURY	7439-97-6
MERCURY COMPOUNDS	EDF-033
MERPHALAN	531-76-0
MESTRANOL	72-33-3
METHACYCLINE HYDROCHLORIDE	3963-95-9
METHAM SODIUM	137-42-8
METHANAMINE, N-METHYL-N-NITROSO	62-75-9
METHAZOLE	20354-26-1
METHIMAZOLE	60-56-0
METHOTREXATE	59-05-2
METHOTREXATE SODIUM	15475-56-6
5-METHOXYPSORALEN	484-20-8
8-METHOXYPSORALEN WITH ULTRAVIOLET A THERAPY	298-81-7
(+-)-METHYL (1R,2R,3R)-3-HYDROXY-2-((E)-(4RS)-4-HYDROXY-4-METHYL-1-OCTENYL)-5- -OXOCYCLOPENTANEHEPTANOATE	59122-46-2
METHYL BROMIDE	74-83-9
METHYL CARBAMATE	598-55-0
METHYL HYDRAZINE	60-34-4
METHYL IODIDE	74-88-4
METHYL MERCURY	22967-92-6
METHYL MERCURY COMPOUNDS	EDF-127
METHYL METHANESULFONATE	66-27-3
2-METHYL-1,3-BUTADIENE	78-79-5
2-METHYL-1-NITROANTHRAQUINONE (OF UNCERTAIN PURITY)	129-15-7

Table 8.2 Listed chemical health hazards. *(continued from previous page)*

Chemical Name	CAS Registry Number (or EDF Substance ID)
1-METHYL-1-NITROSO-3-NITROGUANIDINE	70-25-7
1-METHYL-2-NITROBENZENE	88-72-2
METHYLAZOXYMETHANOL	590-96-5
METHYLAZOXYMETHANOL ACETATE	592-62-1
3-METHYLCHLORANTHRENE	56-49-5
5-METHYLCHRYSENE	3697-24-3
4,4'-METHYLENEBIS (2-CHLOROANILINE)	101-14-4
4,4'-METHYLENEBIS (N,N-DIMETHYL)BENZENAMINE	101-61-1
4,4'-METHYLENEBIS-DIHYDROCHLORIDE BENZENEMINE	13552-44-8
4,4'-METHYLENEDIANILINE	101-77-9
METHYLEUGENOL	93-15-2
METHYLTESTOSTERONE	58-18-4
METHYLTHIOURACIL	56-04-2
METIRAM	9006-42-2
METRONIDAZOLE	443-48-1
MICHLER'S KETONE	90-94-8
MIDAZOLAM HYDROCHLORIDE	59467-96-8
MINOCYCLINE HYDROCHLORIDE (INTERNAL USE)	13614-98-7
MIREX	2385-85-5
MITOMYCIN C	50-07-7
MITOXANTRONE HYDROCHLORIDE	70476-82-3
MONOCROTALINE	315-22-0
5-(MORPHOLINOMETHYL)-3-([5-NITRO-FURFURYLIDENE]-AMINO)-2-OXALOLIDINONE	139-91-3
MUSTARD GAS	505-60-2
MX (3-CHLORO-4-(DICHLOROMETHYL)-5-HYDROXY-2(5H)-FURANONE)	77439-76-0
MYCLOBUTANIL	88671-89-0
N,N-BIS (2-CHLOROETHYL)-2-NAPHTHYLAMINE (CHLORNAPAZINE)	494-03-1
N-ETHYL-N-NITROSOUREA	759-73-9
N-METHYL-2-PYRROLIDONE	872-50-4
N-METHYLOLACRYLAMIDE	924-42-5
N-NITROSO-N-METHYLUREA	684-93-5
N-NITROSO-N-METHYLURETHANE	615-53-2
N-NITROSODI-N-BUTYLAMINE	924-16-3
N-NITROSODIETHANOLAMINE	1116-54-7
N-NITROSODIETHYLAMINE	55-18-5
N-NITROSODIPHENYLAMINE	86-30-6
4-(N-NITROSOMETHYLAMINO)-1-(3-PYRIDYL)-1-BUTANONE	64091-91-4

Table 8.2 Listed chemical health hazards. *(continued from previous page)*

Chemical Name	CAS Registry Number (or EDF Substance ID)
3-(N-NITROSOMETHYLAMINO)PROPIONITRILE	60153-49-3
N-NITROSOMETHYLETHYLAMINE	10595-95-6
N-NITROSOMETHYLVINYLAMINE	4549-40-0
N-NITROSOMORPHOLINE	59-89-2
N-NITROSONORNICOTINE	16543-55-8
N-NITROSOPIPERIDINE	100-75-4
N-NITROSOPYRROLIDINE	930-55-2
N-NITROSOSARCOSINE	13256-22-9
N-(4-[5-NITRO-2-FURYL]-2-THIAZOLYL) ACETAMIDE	531-82-8
NABAM	142-59-6
NAFARELIN ACETATE	86220-42-0
NAFENOPIN	3771-19-5
NALIDIXIC ACID	389-08-2
NAPHTHALENE	91-20-3
NEOMYCIN SULFATE	1405-10-3
NETILMICIN SULFATE	56391-57-2
NICKEL	7440-02-0
NICKEL (II) HYDROXIDE	12054-48-7
NICKEL ACETATE	373-02-4
NICKEL CARBONATE	3333-67-3
NICKEL CARBONYL	13463-39-3
NICKEL COMPOUNDS	NDB000
NICKEL OXIDE	1313-99-1
NICKEL REFINERY DUST	1146
NICKEL SUBSULFIDE	12035-72-2
NICKELOCENE	1271-28-9
NICOTINE AND SALTS	54-11-5
NIFEDIPINE	21829-25-4
NIMODIPINE	66085-59-4
NIRIDAZOLE	61-57-4
NITRAPYRIN	1929-82-4
NITRILOTRIACETIC ACID	139-13-9
NITRILOTRIACETIC ACID, TRISODIUM SALT MONOHYDRATE	18662-53-8
5-NITRO-O-ANISIDINE	99-59-2
5-NITROACENAPHTHANE	602-87-9
2-NITROANISOLE	91-23-6
NITROBENZENE	98-95-3
4-NITROBIPHENYL	92-93-3
6-NITROCHRYSENE	7496-02-8
NITROFEN	1836-75-5

Table 8.2   Listed chemical health hazards. *(continued from previous page)*

Chemical Name	CAS Registry Number (or EDF Substance ID)
2-NITROFLOURENE	607-57-8
NITROFURANTOIN	67-20-9
NITROFURAZONE	59-87-0
1-([5-NITROFURFURYLIDENE]-AMINO)-2-IMIDAZOLIDINONE	555-84-0
NITROGEN MUSTARD HYDROCHLORIDE	55-86-7
NITROGEN MUSTARD N-OXIDE	126-85-2
NITROGEN MUSTARD N-OXIDE HYDROCHLORIDE	302-70-5
NITROMETHANE	75-52-5
2-NITROPROPANE	79-46-9
4-NITROPYRENE	57835-92-4
1-NITROPYRENE	5522-43-0
NORETHISTERONE	68-22-4
NORETHISTERONE (NORETHINDRONE)/ETHINYL ESTRADIOL	EDF-075
NORETHISTERONE (NORETHINDRONE)/MESTRANOL	EDF-074
NORETHISTERONE ACETATE (NORETHINDRONE ACETATE)	51-98-9
NORETHYNODREL	68-23-5
NORGESTREL	6533-00-2
O,P'-DDT	789-02-6
O-ANISIDINE	90-04-0
O-ANISIDINE HYDROCHLORIDE	134-29-2
O-DINITROBENZENE	528-29-0
O-PHENYLENEDIAMINE	95-54-5
O-PHENYLPHENATE, SODIUM	132-27-4
O-TOLUIDINE	95-53-4
O-TOLUIDINE HYDROCHLORIDE	636-21-5
OCHRATOXIN A	303-47-9
OIL ORANGE SS	2646-17-5
ORAL CONTRACEPTIVES, COMBINED	EDF-039
ORAL CONTRACEPTIVES, SEQUENTIAL	EDF-040
OREGON ERIONITE	12510-42-8
OXAZEPAM	604-75-1
OXYDEMETON METHYL	301-12-2
OXYDIAZON	19666-30-9
OXYMETHOLONE	434-07-1
OXYTETRACYCLINE	79-57-2
OXYTETRACYCLINE HYDROCHLORIDE (INTERNAL USE)	2058-46-0
P-A,A,A-TETRACHLOROTOLUENE	5216-25-1
P-CHLORO-O-TOLUIDINE	95-69-2

Table 8.2 Listed chemical health hazards. *(continued from previous page)*

Chemical Name	CAS Registry Number (or EDF Substance ID)
P-CHLORO-O-TOLUIDINE, STRONG ACID SALTS OF	EDF-308
P-CHLOROANILINE	106-47-8
P-CHLOROANILINE.HCL	20265-96-7
P-CRESIDINE	120-71-8
P-DINITROBENZENE	100-25-4
P-NITROCHLOROBENZENE	100-00-5
P-NITROSODIPHENYLAMINE	156-10-5
PACLITAXEL	33069-62-4
PANFURAN S	794-93-4
PARAMETHADIONE	115-67-3
PENICILLAMINE	52-67-5
PENTACHLOROPHENOL	87-86-5
PENTOBARBITAL SODIUM	57-33-0
PENTOSTATIN	53910-25-1
PHENACEMIDE	63-98-9
PHENACETIN	62-44-2
PHENAZOPYRIDINE	94-78-0
PHENAZOPYRIDINE HYDROCHLORIDE	136-40-3
PHENESTERIN	3546-10-9
PHENOBARBITAL	50-06-6
PHENOLPHTHALEIN	77-09-8
PHENOXYBENZAMINE	59-96-1
PHENOXYBENZAMINE HYDROCHLORIDE	63-92-3
PHENPROCOUMON	435-97-2
PHENYL GLYCIDYL ETHER	122-60-1
PHENYLHYDRAZINE	100-63-0
2-PHENYLPHENOL	90-43-7
PHENYTOIN	57-41-0
PHIP (2-AMINO-1-METHYL-6-PHENYLIMIDAZOL) (4,5-B) PYRIDINE	105650-23-5
PIMOZIDE	2062-78-4
PIPOBROMAN	54-91-1
PLICAMYCIN	18378-89-7
POLYBROMINATED BIPHENYLS	PJL335
POLYCHLORINATED BIPHENYLS	1336-36-3
POLYCHLORINATED DIBENZO-P-DIOXINS	PCDD
POLYGEENAN	53973-98-1
PONCEAU 3R	3564-09-8
POTASSIUM BROMATE	7758-01-2
POTASSIUM DIMETHYLDITHIOCARBAMATE	128-03-0
PRAVASTATIN SODIUM	81131-70-6

Table 8.2   Listed chemical health hazards. *(continued from previous page)*

Chemical Name	CAS Registry Number (or EDF Substance ID)
PREDNISOLONE SODIUM PHOSPHATE	125-02-0
PRIMIDONE	125-33-7
PROCARBAZINE	671-16-9
PROCARBAZINE HYDROCHLORIDE	366-70-1
PROCYMIDONE	32809-16-8
PROGESTERONE	57-83-0
PRONAMIDE	23950-58-5
PROPACHLOR	1918-16-7
PROPANE SULTONE	1120-71-4
PROPARGITE	2312-35-8
PROPYLENE GLYCOL BUTYL ETHER	57018-52-7
PROPYLENE OXIDE	75-56-9
PROPYLENEIMINE	75-55-8
PROPYLTHIOURACIL	51-52-5
PYRIDINE	110-86-1
PYRIMETHAMINE	58-14-0
QUAZEPAM	36735-22-5
QUINOLINE AND ITS STRONG ACID SALTS	EDF-217
QUIZALOFOP-ETHYL	76578-14-8
RADIONUCLIDES	1165
RESERPINE	50-55-5
RESIDUAL (HEAVY) FUEL OILS	EDF-049
RESMETHRIN	10453-86-8
RETINOL / RETINYL ESTERS, WHEN IN DAILY DOSAGE IN EXCESS OF 10,000 IU,OR 3,000 RETINOL EQUIVALENTS	68-26-8
RIBAVIRIN	36791-04-5
RIFAMPICIN	13292-46-1
SAFROLE	94-59-7
SALICYLAZOSULFAPYRIDINE	599-79-1
SECOBARBITAL SODIUM	309-43-3
SELENIUM SULFIDE	7446-34-6
SERMORELIN ACETATE	EDF-373
SHALE-OILS	68308-34-9
SILICA	1175
SODIUM DIMETHYLDITHIOCARBAMATE	128-04-1
SOOTS, TARS, AND CERTAIN MINERAL OILS	EDF-050
SPIRONOLACTONE	52-01-7
STANOZOLOL	10418-03-8
STERIGMATOCYSTIN	10048-13-2
STREPTOMYCIN SULFATE	3810-74-0
STREPTOZOTOCIN	18883-66-4

Table 8.2 Listed chemical health hazards. *(continued from previous page)*

Chemical Name	CAS Registry Number (or EDF Substance ID)
STRONG INORGANIC ACID MISTS CONTAINING SULFURIC ACID	7664-93-9(InorgMist)
STYRENE OXIDE	96-09-3
SULFALLATE	95-06-7
SULINDAC	38194-50-2
TALC CONTAINING ASBESTIFORM FIBERS	1190
TAMOXIFEN AND ITS SALTS	10540-29-1
TAMOXIFEN CITRATE	54965-24-1
TECHNICAL GRADE 2,4 & 2,6 DINITROTOLUENE	EDF-210
TEMAZEPAM	846-50-4
TENIPOSIDE	29767-20-2
TERBACIL	5902-51-2
TERRAZOLE	2593-15-9
TESTOSTERONE AND ITS ESTERS	58-22-0
TESTOSTERONE CYPIONATE	58-20-8
TESTOSTERONE ENANTHATE	315-37-7
2,3,7,8-TETRACHLORODIBENZO-P-DIOXIN (TCDD)	1746-01-6
1,1,2,2-TETRACHLOROETHANE	79-34-5
TETRACHLOROETHYLENE	127-18-4
TETRACYCLINE (INTERNAL USE)	60-54-8
TETRACYCLINE HYDROCHLORIDE	64-75-5
TETRACYCLINES	EDF-052
1,1,2,2-TETRAFLUOROETHYLENE	116-14-3
TETRANITROMETHANE	509-14-8
THALIDOMIDE	50-35-1
THIOACETAMIDE	62-55-5
THIODICARB	59669-26-0
THIOGUANINE	154-42-7
THIOPHANATE-METHYL	23564-05-8
THIOURACIL	141-90-2
THIOUREA	62-56-6
THORIUM DIOXIDE	1314-20-1
TOBACCO SMOKE (PRIMARY)	EDF-073
TOBACCO, ORAL USE OF SMOKELESS PRODUCTS	EDF-053
TOBRAMYCIN SULFATE	49842-07-1
TOLUENE	108-88-3
TOLUENE DIISOCYANATE (MIXED ISOMERS)	26471-62-5
TRANS-2-([DIMETHYLAMINO] METHYLIMINO)-5- (2-5-NITRO-2-FURYL)VINYL-1,3,4-OXADIAZOLE	55738-54-0
TREOSULFAN	299-75-2
TRIADIMEFON	43121-43-3

Table 8.2   Listed chemical health hazards. *(continued from previous page)*

Chemical Name	CAS Registry Number (or EDF Substance ID)
TRIAZIQUONE	68-76-8
TRIAZOLAM	28911-01-5
TRIBROMOMETHANE	75-25-2
TRIBUTYLTIN METHACRYLATE	2155-70-6
TRICHLORMETHINE (TRIMUSTINE HYDROCHLORIDE)	817-09-4
1,1,2-TRICHLOROETHANE	79-00-5
TRICHLOROETHYLENE	79-01-6
2,4,6-TRICHLOROPHENOL	88-06-2
1,2,3-TRICHLOROPROPANE	96-18-4
TRICYCLOHEXYLTIN HYDROXIDE	13121-70-5
TRIENTINE HYDROCHLORIDE	38260-01-4
TRIFORINE	26644-46-2
TRILOSTANE	13647-35-3
TRIMETHADIONE	127-48-0
TRIMETHYL PHOSPHATE	512-56-1
2,4,5-TRIMETHYLANILINE AND ITS STRONG ACID SALTS	EDF-218
TRIMETREXATE GLUCURONATE	82952-64-5
TRIPHENYLTIN HYDROXIDE	76-87-9
TRIS (1-AZIRIDINYL) PHOSPHINE SULFIDE (THIOTEPA)	52-24-4
TRIS (2,3-DIBROMOPROPYL) PHOSPHATE	126-72-7
TRIS (2-CHLOROETHYL) PHOSPHATE	115-96-8
TRP-P-1 (TRYPTOPHAN-P-1)	62450-06-0
TRP-P-2 (TRYPTOPHAN-P-2)	62450-07-1
TRYPAN BLUE	72-57-1
UNLEADED GASOLINE (WHOLLY VAPORIZED)	GCE100
URACIL MUSTARD	66-75-1
URETHANE	51-79-6
UROFOLLITROPIN	97048-13-0
VALPROATE	99-66-1
VINBLASTINE SULFATE	143-67-9
VINCLOZOLIN	50471-44-8
VINCRISTINE SULFATE	2068-78-2
VINYL BROMIDE	593-60-2
VINYL CHLORIDE	75-01-4
4-VINYLCYCLOHEXENE	100-40-3
WARFARIN AND SALTS	81-81-2
2,6-XYLIDINE	87-62-7
ZILEUTON	111406-87-2

(Source: California EPA and USEPA 2004

Table 8.3 OSHA HAZCOMM definitions. *(continued on next page)*

1. "Carcinogen:" A chemical is considered to be a carcinogen if:

   (a) It has been evaluated by the International Agency for Research on Cancer (IARC), and found to be a carcinogen or potential carcinogen; or

   (b) It is listed as a carcinogen or potential carcinogen in the Annual Report on Carcinogens published by the National Toxicology Program (NTP) (latest edition); or,

   (c) It is regulated by OSHA as a carcinogen.

2. "Corrosive:" A chemical that causes visible destruction of, or irreversible alterations in, living tissue by chemical action at the site of contact. For example, a chemical is considered to be corrosive if, when tested on the intact skin of albino rabbits by the method described by the U.S. Department of Transportation in appendix A to 49 CFR part 173, it destroys or changes irreversibly the structure of the tissue at the site of contact following an exposure period of four hours. This term shall not refer to action on inanimate surfaces.

3. "Highly toxic:" A chemical falling within any of the following categories:

   (a) A chemical that has a median lethal dose (LD(50)) of 50 milligrams or less per kilogram of body weight when administered orally to albino rats weighing between 200 and 300 grams each.

   (b) A chemical that has a median lethal dose (LD(50)) of 200 milligrams or less per kilogram of body weight when administered by continuous contact for 24 hours (or less if death occurs within 24 hours) with the bare skin of albino rabbits weighing between two and three kilograms each.

   (c) A chemical that has a median lethal concentration (LC(50)) in air of 200 parts per million by volume or less of gas or vapor, or 2 milligrams per liter or less of mist, fume, or dust, when administered by continuous inhalation for one hour (or less if death occurs within one hour) to albino rats weighing between 200 and 300 grams each.

4. "Irritant:" A chemical, which is not corrosive, but which causes a reversible inflammatory effect on living tissue by chemical action at the site of contact. A chemical is a skin irritant if, when tested on the intact skin of albino rabbits by the methods of 16 CFR 1500.41 for four hours exposure or by other appropriate techniques, it results in an empirical score of five or more. A chemical is an eye irritant if so determined under the procedure listed in 16 CFR 1500.42 or other appropriate techniques.

5. "Sensitizer:" A chemical that causes a substantial proportion of exposed people or animals to develop an allergic reaction in normal tissue after repeated exposure to the chemical.

6. "Toxic." A chemical falling within any of the following categories:

   (a) A chemical that has a median lethal dose (LD(50)) of more than 50 milligrams per kilogram but not more than 500 milligrams per kilogram of body weight when administered orally to albino rats weighing between 200 and 300 grams each.

Table 8.3 OSHA HAZCOMM definitions. *(continued from previous page)*

(b) A chemical that has a median lethal dose (LD(50)) of more than 200 milligrams per kilogram but not more than 1,000 milligrams per kilogram of body weight when administered by continuous contact for 24 hours (or less if death occurs within 24 hours) with the bare skin of albino rabbits weighing between two and three kilograms each.

(c) A chemical that has a median lethal concentration (LC(50)) in air of more than 200 parts per million but not more than 2,000 parts per million by volume of gas or vapor, or more than two milligrams per liter but not more than 20 milligrams per liter of mist, fume, or dust, when administered by continuous inhalation for one hour (or less if death occurs within one hour) to albino rats weighing between 200 and 300 grams each.

7. "Target organ effects."

The following is a target organ categorization of effects which may occur, including examples of signs and symptoms and chemicals which have been found to cause such effects. These examples are presented to illustrate the range and diversity of effects and hazards found in the workplace, and the broad scope employers must consider in this area, but are not intended to be all-inclusive.

(a) Hepatotoxins: Chemicals which produce liver damage

Signs & Symptoms: Jaundice; liver enlargement

Chemicals: Carbon tetrachloride; nitrosamines

(b) Nephrotoxins: Chemicals which produce kidney damage

Signs & Symptoms: Edema; proteinuria

Chemicals: Halogenated hydrocarbons; uranium

(c) Neurotoxins: Chemicals which produce their primary toxic effects on the nervous system

Chapter 9

BIOLOGICAL HAZARDS

CHARACTERIZATION OF BIOHAZARDS

Biohazard sources may include micro-organisms, insects, plants, animals, birds, and humans. Biohazard-induced human maladies could range from allergy, irritation, infection, toxicity, cellular mutation, cancer, bio-physical disorders, tissue injuries, and others:

Properties of microbial biohazards

- For purposes of this text, a microbiological agent that is capable of causing disease in people and/or animals, or exhibits a potential to cause negative effects upon the environment will be considered a biohazard. In this regard we will classify microbial agents (microbes) inclusive of viruses, bacteria, fungi, and similar parasites that are capable of causing disease; or that pose a threat to the environment, to be *pathogens*. We will also point out that some microbes require oxygen to exist, others require carbon dioxide, and others can make do wherever that may reside, *facultative*.
- What about hazardous conditions caused by microbes? For example, hazardous by-products that are produced by microbiological agents should also be considered biohazards. Examples would include: botulism toxins, staphylococcal toxins, fungal myco-toxins and many, many others.

Further Characterization

Along with specific physical-biological properties, biohazards should be classified for their potential effects upon normal life patterns in humans and upon the surrounding environment. The following should be known about a potential biohazard:

1. Is the organism capable of causing infection, illness, and disease in humans?
2. Could it have a negative effect upon the environment, especially in the food chain?
3. Is there a reservoir or growth source for the microbe that would allow it to grow and proliferate?
4. What are the potentials for a given pathogen to be transferred from its growth source and to be spread into an unsuspecting population?
5. What is the etiology or ways that a pathogenic organism can be transferred throughout a given population?
6. What media may be affected by contamination from the organism e.g. water, food, air quality, and others?
7. How may a given population be affected when exposed to a given biohazard, e.g. susceptibility, immunity, toxicity effects and others?
8. Is there a way to interrupt the progress of contamination or infection—can exposure to the biohazard and its specific etiology be controlled?

Affected Populations

The disasters that occurred during and after the 2004 Christmas Vacation Tsunami in Southeast Asia are frightening examples of how vulnerable unsuspecting populations can be when biohazards become uncontrolled. In the aftermath of destruction from horrendous forces of water damage, thousands of humans, animals, and plant life perished and many openly decayed within the affected environment. Widespread microbial infections and all varieties of contamination spread throughout the disaster sites. This calamity was the result of a natural disaster; however similar abominations would also occur from the inhumane acts of mankind.

OCCUPATIONAL EXPOSURE TO BIOHAZARDS

There are many workers that have a potential for exposure to biological hazards, for example:

1. Agriculture Workers
   - Personnel involved in cultivating and harvesting crops, especially those involved in natural fertilization
   - Workers involved with breeding and tending animals

- Forestry workers
- Fish and game workers

2. Handlers of Agricultural Products
   - Animal meat preparation and meat processing
   - Operators of grain processing facilities
   - Processors for tanning and animal skin products
   - Producers of textiles
   - Processors of paper, wood, and cork

3. Laboratory Workers
   - Microbiology lab personnel
   - Research, technology, and animal testing personnel
   - Pharmaceutical producers
   - Mortuary workers

4. Health Care Professionals
   - Medical treatment providers
   - Dental treatment providers
   - Medical laboratory and associated personnel
   - Hospitals and related services personnel
   - Veterinarian services providers

5. Personal Care Providers
   - Hair care service personnel
   - Skin and beauty treatment personnel
   - Tattoo parlor operators
   - Daycare and child care providers

6. Environmental, Health, Safety, and Emergency Service Personnel
   - Environmental, health and safety (EHS) management personnel
   - EHS trainers, instructors, and professional services providers
   - Environmental hygiene and industrial hygiene personnel
   - Police, Crime Scene Investigation (CSI) and forensics personnel
   - On-scene EMS providers, first aid, and ambulance personnel
   - Fire services and fire investigators
   - HAZMAT emergency response team personnel
   - Disaster site workers
   - Decontamination contractors
   - Military services personnel

7. Site Remediation & Clean Up Personnel
   - Fire damage clean up contractors
   - Water damage clean up and repair contractors
   - Mold and water borne microbial decontamination personnel

8. Building Services and Facility Services Personnel
   - Building maintenance workers
   - Janitorial services workers
   - Heating, ventilating, and air conditioning personnel (HVAC)

9. Recreational & Sports Workers
   - Pool, spa, and exercise center workers
   - Coaches, referees, and athletic support workers
   - Room clean up personnel in hotels and lodging services

10. Sanitary Services
    - Garbage and trash collection personnel
    - Sewer treatment, publicly owned and private sewage treatment workers
    - Wastewater operators
    - Non-regulated waste treatment, storage, and disposal personnel
    - Regulated waste, universal waste, and special waste site workers

11. Community Services Personnel
    - Homeless care and shelter providers
    - Teachers and educators
    - Social services personnel

We would be remiss without considering specific examples of biological hazards that may be encountered when working in some careers. First of all, don't forget the four prime methods that people become exposed to hazardous substances, including biohazards:

- Inhalation which is usually the quickest mode of exposure of infected aerosols, mists, dusts, and airborne contaminated particles
- Entry into holes and cracks in the skin (including puncture wounds from sharp contaminated objects)
- Ingested or swallowed contaminated materials, including materials that are collected into the nasal mucosa and are eventually swallowed. Remember that Mama told you to wash your hands before eating? There is a reason to do that.

- Absorption through the skin, eyes, and other tissues; hazardous chemicals produced by microbial agents can be absorbed through the skin and topical skin infections often occur.

Virus Exposures

Where would we most likely be exposed to viral biohazards? Remember that viral microbial agents are obligate parasites; in other words, they usually require living cells to remain active. Viral pathogens are so small that they penetrate cells found in all circulating liquids and tissues within the body. They cannot be seen without extreme magnification such as that seen using electron microscopy and more sophisticated methods. Viruses are not considered to be complete cells; in this regard their response to antibiotic treatment is limited.

- Any occupation that workers may come in contact with blood or body fluids must assume that potentials of virus contamination are very possible. This opens the door for many of the above listed workers to beware and protect themselves from Aids, Hepatitis, and many other bloodborne viral agents.
- Workers in occupations that could be bitten by animals and insects must also know that viral infection is potential e.g. mosquitoes are vectors (carriers) of the West Nile virus, Viral Meningitis and other diseases. Several animals are vectors for the Rabies virus and several other pathogenic viral agents.

Bacterial Exposures

Bacteria are larger than viruses and they have cellular morphology, in this regard they usually respond to antibiotic treatment—unless they have become antibiotic resistant. Bacteria can be seen using conventional "low powered" microscopy, and bacterial morphologies or shapes are quite simple—*rods and circles in clumps, chains, or as individuals.*

Please note that many bacteria also exist as or form into "spores" and properties of spores add to increased exposure potential. What is a spore? A spore is actually a miniature suit of armor that is worn by the micro-organism. A microbe that can form into a spore is capable of withstanding environmental changes such as extremes of temperature, lack of moisture, and chemical exposure. Biological weapons of mass destruction such as anthrax, botulism and many others are spore formers. Note that spores appear dry and dusty and they can easily enter the air that we breathe.

Bacterial infections can occur as local infection of the skin, respiratory system or those having more systemic effects within the body. Occupations that have a propensity for exposure to bacterial pathogens include the following:

- Anyone that may be subjected to infected blood or body fluids

- Workers that are involved with health care and laboratory services, veterinary and handling of animals, food processing, waste handling and treatment, wood and paper processing, water damage remediation, emergency action and response, personal and beauty services, building services and maintenance, construction, outdoor activities and others.

Fungal Exposures

Fungi are enemies with bacteria and they compete with one another for food sources, in this regard there is a zone of demarcation between them in environments conducive to allowing each to grow. Note that the growth of fungi and bacteria are on opposite sides of the media. Fungi produce myco-toxins that kill bacteria and bacteria produce bacterial-toxins that kill fungi. The properties of microbial toxins are used for producing drugs such as antibiotics and, in some situations, they are used for producing weapons of mass destruction.

Fungi are found in every world environment, especially in the spore form. They love moist "watery" environments and enjoy food sources containing sugar, irregardless if it is contained in animal or human foods, or in organic dusts that blow throughout the atmosphere. They thrive in areas with subdued light to darkness. They can produce health concerns in most humans, ranging from allergies, local skin infections, respiratory infections, subcutaneous infections, and systemic health conditions. Occupations that are often exposed to fungal health hazards include:

- Occupations that are subject to humid, watery conditions. Plumbers, confined space workers, construction, building engineering and facilities maintenance, heating-ventilating-air conditioning (HVAC) personnel, and water treatment personnel.
- Personnel involved in excavating, mining, sewer and wastewater treatment, fire and water damage remediation, and mold abatement can be exposed to fungal biohazards.

Pathogenic Parasites

This grouping includes pathogens found in contaminated water, soils, and pathogens that are transmitted by insects and directly or indirectly from wild animals. Parasitic diseases pose health concerns that are potential for personnel that work outdoors, especially in the wilds. Often personnel that work in sand and sandy soils that are contaminated with human or animal feces have a high incidence of parasitic infection. The most common disease causing parasites are protozoa, helminthes, and arthropods:

- Protozoan diseases are transmitted by contaminated food, water, direct contact, and from insect bites (vectors). Examples of diseases caused by protozoa infections include: Malaria, Amebic Dysentery, Enteritis, Vaginitis, Sleeping Sickness, Blood and Tissue Disease (Kala-azar), and others.

- Helminthes, or parasitic worms, cause disease using similar mechanisms as protozoa. Disease in humans includes: Elephantiasis, Trichinosis, Tapeworm, Hookworm, Bleeding Dysentery, Skin Diseases, and others.
- Arthropod vector bites (mosquitoes, ticks, chiggers, etc), wild animals, and birds transmit pathogenic diseases as they are infected from bacteria, rickettsiae, viruses, and protozoa. Diseases caused from these sources include: Bubonic Plague, Septicemia and Pneumonic Plagues, Tularemia, Yellow Fever, Dengue Fever, Epidemic Encephalitis, Malaria, Sleeping Sickness, Spotted Fever, Lime Disease, Typhus, Rickettsialpox, Trench Fever, and others. Note: Anthrax bacteria are carried by sheep, goats, deer, and other animals; and Rabies virus is carried by rodents (including bats) and bites of several types of infected animals.

SELECTED BIOHAZARD SPECIFICS

The U.S. Occupational Safety and Health Administration (OSHA), Centers for Disease Control (CDC) and the National Institute of Occupational Safety and Health (NIOSH) lists the following biological hazard concerns from perspectives of occupational safety and for considering potential threats of bioterrorism. Reference: OSHA.gov/Biological Agents, 12/2004.

As we have previously discussed, bacteria, viruses, fungi, and other micro-organisms, along with their associated toxic products, are considered to be biological agents. The hazards associated with biological agents (biohazards) have the potential to affect human health. Again as previously mentioned, human health effects related to biohazards can range from mild reactions to death—depending upon amounts of exposure and the condition of a victim's immune system. Because most microbial agents rapidly reproduce and often require only limited resources to survive, they pose significant danger and are characterized as living hazardous materials.

The reader should try to remember the following biological agents as your career may take you into the hazardous materials career fields:

Anthrax

After 9/11, every person that lives in the free world has heard something about the Bacillus Anthracis or Anthrax. This biological agent has the properties of a spore-forming bacterial organism (bacterium). One form of disease caused by anthrax is called "wool sorters disease" because it is usually acquired through having contact with anthrax-infected animals, or from contacting contaminated animal bi-products. Because it is relatively easy to obtain and easy to disseminate, it has been used as a biological terrorism agent.

Anthrax infections usually occur by three mechanisms of exposure:

- Skin contact or cutaneous disease, if not properly treated, will cause major degradation of the skin and become a deadly systemic disease

- Inhalation of anthrax spores which causes a respiratory infection that proliferates quickly and yields a significant human death rate
- If contaminated products are ingested, or if undercooked infected meat is eaten, anthrax may cause disease in the gastrointestinal system, which also can yield a significant human death rate. It should be mentioned that it is rare to find any anthrax-infected animals in the United States.

Note: Anthrax infection caused by person-to-person contact is unlikely. Also, dealing with patients that have anthrax respiratory infections do not pose unreasonable risks for health care workers. Unfortunately, media attention conducted in the wake of the September 11, 2001 terrorist attacks on the United States appears to have incorrectly described anthrax etiology.

There is an immunization for anthrax; however it *does not work* for 100% of persons that are vaccinated. Due to increased potentials for exposure, the *Advisory Committee on Immunization Practices* has recommended that anthrax vaccinations should be provided to the following workers: (reference: CDC Advisory Committee):

- Laboratory workers that handle anthrax during testing and research
- Workers that process animal skins or fur (tanneries, wool mills, etc.)
- Veterinarians and associated workers that may deal with infected animals
- Military personnel that serve in locations that may pose risks of using anthrax as a bio-terrorism or warfare agent
- Emergency responders and decontamination personnel that may handle incidents involving anthrax contamination

Influenza vs. Avian Flu

Most of the world is routinely affected by influenza which is a disease that is characterized by fever, muscle pain, and respiratory distress. It is caused by a variety of viruses and the severity of the disease depends upon a patient's medical condition and specific properties of the virus.

Influenza viruses are characterized using a 3-part classifying system that indicates:

- Type of virus: A, B, C
- Sub-type: Classification is based upon surface coatings on the virus, which indicate whether the virus will affect humans and other kinds of animals.
- Strain: The scientists further classify a virus's potential effect upon humans and animals based upon genetic differences noted between specific types and sub-types of influenza. This is one of the challenges that must be addressed by the makers of flu shots—if they forecast "guess" the most probable strains that will affect a given population, the shot may work. Sometimes a virus that is affecting humans or animals in one part of the world is not the same as one that is causing the same affects in other parts of the world.

Avian "Bird Virus" Genetic Variation

Influenza viruses will mutate or change over a given period of time. For instance, if a human is infected by a common human influenza virus and a "bird virus" (avian type virus) the two viruses could exchange genetic properties. In this regard, viral gene trading inside a human body may develop a new subtype of influenza virus that most people would not have any immunity. Because of this, the bird virus could be transferred from birds to people and from person to person. The current avian virus is thought to have limited person to person transfer, and is primarily transferred from birds to people from infected poultry or from contaminated materials. At the time of this writing, the Avian Influenza type A Virus, strain H5N1, is considered by CDC to be highly pathogenic.

Blood Borne Pathogens

As we drive down the highway we may witness a terrible accident. A thought then comes to mind, "Should I stop to help and take a chance on being exposed to Aids or hepatitis?" It wasn't very long ago that everyone would stop and help the injured bleeding victims; unfortunately today we live in a different world.

OSHA has implemented directives that are to be followed whenever human blood or body fluids are required to be handled in the work place, due to potentials for exposure to blood borne pathogens. What is considered to be a blood borne pathogen? Common sense would direct that anything emitted from a diseased body should be handled using universal precautions. This author still honors that philosophy. The OSHA Standard focuses upon viral infections that are difficult to treat or that are fatal. The viruses include:

- Human Immunodeficiency Virus (HIV) that yields AIDS.
- Hepatitis B Virus
- Hepatitis C Virus and other non-treatable viruses that may follow

Who has the highest potential to be exposed to blood borne pathogens (BBP)?

1. Anyone that may be exposed to human blood or body fluids
2. Health care, medical and dental support personnel
3. Emergency action, first aid, and emergency response personnel
4. Disaster site workers and incident recovery workers
5. Public safety, police, fire services, and crime scene investigators
6. Janitorial and facility services personnel

How may workers be protected when dealing with BBP? By following the same procedures that anyone needs to protect themselves from biohazards—*by following universal precautions.*

What are universal precautions? The answer is simple, by keeping contaminants from infecting your body. How? Remember in previous chapters we discussed using chemical protective equipment and respiratory protection? Don't forget that microbes can attack the human body by the same modes of exposure as chemicals. We do not want pathogenic microbes on our clothing, on our skin, or entering into our bodies. For the most part, BBP microbes enter the body via sharp objects, cracks in the skin, through mucous membranes in the mouth, eyes, respiratory system, genitalia, and others. When providing treatment to victims whenever blood or body fluids are involved, the following should be considered:

- Use procedures that will not allow sharp objects to puncture your skin.
- Use goggles or eye shields
- Protect respiratory entry points as necessary (dust mask, respirator)
- Cover bare skin (ideally with non-penetrating clothing, rubber, or plastic)
- Use disposable rubber gloves

AIDS

What is it? Acquired Immune Deficiency Syndrome (AIDS) and it means that a victim that has been infected by the human immunodeficiency virus (HIV) no longer has a healthy immune system. The condition that is acquired from the HI virus causes victims to suffer unto death from diseases that a normal immune system would resist.

How can this happen? The body's immune system relies upon cooperative efforts between white blood cells (WBC's) that fight invading micro-organisms and contaminants. The key defensive cells are designated as B-cells and T-cells.

The B-cells produce proteins, called antibodies, that bind to inactivate an invading organism. After the immune system has been exposed to an invader, the circulating antibodies enhance the body's immune-memory system. The strength of the antibody memory system determines the body's ability to keep from having return attacks from remaining pathogens. A healthy immune system will contain antibodies that will recognize a pathogen upon its return visit, and inactivate it. T-cells are the hawks of the immune system; they directly attack invading pathogens and kill them.

There are also some white blood cells that are the *special forces of the immune system*; they are called T-helper cells. The purpose of these cells is to stimulate or awaken the B-cells and T-cells and direct them to attack invading pathogens. They all work together to methodically wipe out microbiological invaders in the human bloodstream and within the tissues.

How can HIV cause AIDS if there is so much antibody protection? The Human Immunodeficiency Virus is small and has some special chemical properties about it that actually turns a T-helper cell into a Trojan Horse. The HIV penetrates T-helper cells and once it gets inside the T-helper cell, it multiplies. The replicating HIV produces toxins that kill the T-helper cell. When this activity occurs in significant numbers, the normal immune system becomes crippled and circulating B-cells and T-cells are not activated to attack the simplest of daily invading pathogens. Unfortunately, victims of AIDS die from diseases that people with normal immune systems treat with cold medicine.

How is the cause of AIDS transmitted? HIV can be normally transferred from person to person by the following mechanisms:

- Sharing sexual activities with an infected person
- Puncture wounds from an infected sharp object
- Transfusions of infected blood
- Transfusions of infected blood products
- Infection of an unborn child from an HIV infected mother
- Infection from receiving organ transplants from an infected donor.
- Not taking universal precautions when they are required

"We cannot solve our problems with the same level of thinking that created them," by Albert Einstein.

Botulism & Genus Clostridium

Botulism is a disease that is generated by toxins produced by a spore-forming bacteria that belongs to a family of microbes that produce extremely hazardous toxins.

About Bacterial Spores

As mentioned earlier in this chapter, microbial spores are actually suits of armor that are formed to protect the microbial agent from the environment. Bacterial spores are found almost everywhere. They can ride fine, almost invisible, particles of dust and the air moves them throughout the environment. They are hard to destroy and high temperatures are usually required to kill them. Clostridium spores are plentiful outdoors, in the soil, and on foods in the garden. The pressure cooker was invented for use in safely canning food, especially low acid to bland foods because Clostridium botulinum loves to hatch from a spore and replicate in a low oxygen environment common to canned food.

The family, or genus, Clostridium has three members that should be noted by hazardous materials personnel:

1. Clostridium tetani: This organism produces the disease called tetanus, also called lockjaw. Victims of this disease usually become infected by having a puncture wound that injects Clostridium tetani spores into the wound and as the spores become active bacterial cells and replicate, they generate a toxin that causes systemic rigid paralysis. Victims that suffer from this disease, that are not treated, can suffer a very painful death. Today, most people have been

vaccinated and should be revaccinated periodically. Whenever a person is treated at a hospital emergency room, the first question usually asked is, "When was your last tetanus shot?"

2. Clostridium perfringens: This organism causes gas gangrene in wounds that are not properly treated. Prior to the advent of antibiotic drugs, it was common to see many amputees that lost a limb because they became infected and gangrene occurred in the wound. If they had a wound that could not be treated by amputation, they usually died.

3. Clostridium botulinum: This organism produces the disease called botulism. C. botulinum as a microbe is not exceptionally pathogenic; however it produces one of the most poisonous substances known. There are three primary kinds of botulism:

   • Food borne botulism: Caused from food that contains botulism toxin. The organism often breaks down canned food (while sealed in the can) and if it is not heated for extended periods of time, the toxin remains extremely potent. In this regard, small amounts of toxin will poison many people. Botulism toxin causes flaccid paralysis, such that victim's muscles will become extremely loose, they cannot swallow and may choke to death on their own saliva. The muscular action of the heart also becomes impeded. Prompt medical treatment is needed to save a victim's life.

   • Wound botulism: This condition is caused by microbial action that produces toxins in an infected wound.

   • Infant botulism: Children may ingest materials that are contaminated by botulinum spores. As they replicate in the child's intestines, the botulism toxin is generated. As previously mentioned, the toxin causes flaccid paralysis in the child.

All forms of botulism may be fatal and prompt medical treatment is required to save the victim's life.

Food Borne Diseases (Food Poisoning)

There are more than 250 different diseases that can occur from infected or toxic contaminated food. Food poisoning can be caused by viruses, bacteria, parasites, microbial toxins, chemical toxins, metals, and other substances.

Diseases may range from stomach upset to life-threatening illnesses that may affect the nervous system, liver, kidneys, and other organs. Some of the more common food borne diseases includes:

• Botulism
• Brucellosis
• Campylobacter enteritis

- Escherichia coli infections
- Hepatitis A
- Listeriosis
- Salmanellosis
- Shigellosis
- Toxoplasmosis
- Viral Gastroenteritis
- Taeniasis
- Trichinosis

Why are food borne diseases significant to health and hazardous materials professionals? Because food borne diseases can produce large numbers of casualties within short durations of time.

Hantavirus

Disease related to this organism primarily is generated from infected rodents. The disease is transmitted from dried rodent feces, urine, or saliva. People that breathe contaminated dusts often become ill. Initial symptoms of Hantavirus infection appear to be flu-like, such that victims have fever, chills, and aching muscles. Unfortunately, the disease rapidly progresses to a severe pneumonia and the lungs fill with fluids—often causing death in humans.

People at risk include:

- People that work in laboratory environments with rodents
- People that work in infected rodent-contaminated buildings
- Construction personnel that work in dusty environments

Legionnaires Disease

Often building maintenance and heating, ventilating, and air conditioning units are infected with the bacteria called Legionnella. Each year 10,000 to 50,000 people are infected with the potentially fatal respiratory infection. Why? They do not protect themselves with proper personal protective equipment, especially respiratory protection. The following areas are common locations for Legionnella and personnel should be required to wear personal protective equipment before entering:

- Locations with poor ventilation, that contain water-based aerosols
- Vents and ducting
- Humid confined spaces
- Air conditioning cooling towers

- Potable water systems (poorly maintained and confined)
- Wastewater treatment systems
- Pits and sump areas associated with fountains, pools, spas, and similar locations
- Processing areas with poor ventilation, high humidity, or watery mists
- Fire-damaged and water-damaged areas
- Disaster sites

Ref: OSHA, Biological Agents

Plague

The plague is well documented throughout human history; millions of people have perished from it. According to the World Health Organization (WHO), there are approximately 3,000 cases of plague globally each year. The plague is caused by a bacillus called Yersinia pestis and it is carried by fleas and other types of vectors. Today conditions have been reduced around the world to prevent another pandemic; however, there are bio-terrorism concerns about releasing pneumonic forms of the disease—even today this action could present death and disaster to any affected population. (Ref: CDC)

Disease symptoms:

- Bubonic plague: Oversized, painful lymph nodes, fever, chills, and total weakness in all bodily functions
- Septicemic plague: High fever, chills, abdominal pain, shock, bleeding into skin and into other organs, and total weakness in all bodily functions
- Pneumonic plague: High fever, chills, respiratory distress, coughing, difficulty breathing, severe shock, and death if untreated

Note: If prompt medical treatment is not provided for plague victims, the death rate is approximately 90%. In cases that receive prompt medical treatment, the death rate is approximately 15%.

Smallpox

The disease smallpox is caused by the Variola Virus, which is extremely contagious and is somewhat unique to humans. The virus is transferred from person to person, from direct contact with infected persons, or by inhalation of air droplets, saliva, or aerosols generated by infected persons.

Smallpox is also a bio-terrorism concern that would have a major effect upon any world population. In most nations, less than 20% of the population has been appropriately vaccinated. Because the disease is primarily transmitted by respiratory exposures, controlling epidemic incident levels would be difficult. There is no complete treatment for smallpox. There would not be enough medical facilities in most nations to properly treat and isolate victims.

Tularemia

The extremely infectious bacteria, Francisella tularensis causes the disease called tularemia. It occurs naturally in the United States and is found in rodents, rabbits, and hares. A unique property of Tularemia is that it only requires few bacteria to cause and propagate the disease; because of this property it is also a bio-terrorism concern.

Disease symptoms include:

- Rapid onset of high fever & chills
- Swollen and painful glands
- Headache, muscle pain, and joint pain
- Diarrhea
- Sore eyes, sore throat, and dry non-productive cough
- Progressive total body weakness
- In some cases, chest pain, pneumonia, trouble breathing, and bloody sputum
- Depending upon mode of exposure, ulceration can occur on the skin and in the mouth

Mechanism of disease transfer:

- Bites from infected insects such as ticks, deerflies, and others
- Handling infected dead animals without proper personal protective equipment
- Ingesting contaminated food or water
- Inhalation of contaminated dusts or aerosols

Viral Hemorrhagic Fevers (VHF)

A group of diseases depicted as VHF are caused by ribonucleic acid (RNA) viruses that are characterized by four distinct families:

1. Ebola Hemorrhagic Fever
2. Marburg Hemorrhagic Fever
3. Lassa Fever
4. Hantavirus Pulmonary Syndrome and Yellow Fever

Each virus characterization produces various symptoms of disease. Most include high fever; total body weakness; pain in muscles and joints. In some cases there is bleeding from several parts of the body, however death is not usually associated with blood loss. In severe cases the victims will fall into shock and coma. Even though some VHF's produce mild illness, many produce severe life-threatening conditions with high death rates.

Note: Hemorrhagic fever viruses are identified by the Centers for Disease Control and Prevention (CDC) as part of six agents that are likely to be used in biological warfare agents.

Chapter 10

DISASTER SITE WORK

We discussed issues related to human health hazards, especially situations that involve dangerous goods and chemicals in accidental releases and at disaster sites. In this chapter we will refer to concerns relating to biological hazards including a major source of contamination and infection, water damage.

You will note that a large portion of this chapter has been dedicated to water damage clean up, also called site remediation. Unfortunately now embedded in our souls, along with biological fears related to bio-terrorism, are horrors related to the catastrophic affects from natural disasters, water damage at an unimaginable magnitude, the 2004 Christmas Vacation Tsunami and Hurricane Katrina, the largest natural disaster to occur on the American continent.

Early in 2004, efforts were made by OSHA and the National Institute of Environmental Health Sciences (NIEHS) by means of the Hazardous Materials Training, and Research Institute (HMTRI) (CCCHST Program) and the OSHA Outreach Training Center Satellites across the United States to train trainers for the purpose of training workers to work safely at disaster sites. We did not know that our disaster site worker program was going to be needed worldwide as well as to respond to major disasters in Mississippi and New Orleans, Louisiana.

Key principles of individual safety that are emphasized in the OSHA Disaster Site Workers Training Program are emphasized in this chapter as well.

DISASTER

Disaster is defined by Webster as "an unforeseen mischance bringing with it destruction of life or property or utter defeat."

Of course, natural disasters may occur anywhere or at anytime. Natural disasters seem to bring with them the human realization that mankind really does not possess powers to control them and enforces that fact that we are only visitors on this planet.

We discuss potentials of disasters that could result from actions of mankind, terrorist activations. Situations that result from terrorism do not meet the above mentioned definition taken from Webster's New Collegiate Dictionary because terrorist actions are not the result of unforeseen actions. In this regard, destruction sites caused by terrorists should be considered crime scenes and in this text they will be indicated as such.

Natural disaster, disaster that occurs from accidental actions, and crime scenes all have one thing in common. They must be stabilized, controlled, and put back to normal so life can go on.

PUTTING DISASTER SITES BACK TO NORMAL

Disaster that involves loss of life, fire, water damage, and large areas of destruction rapidly becomes reservoirs for disease and pollution. Everyone that enters a disaster site will become exposed to dangers found from within. To be able to meet disaster response objectives, all entrants must protect themselves from inherent hazards and must know how to safely perform required duties.

To keep from becoming part of the disaster, all entrants must:

- Have a specific reason or need (be authorized) to enter the disaster site
- Be trained to safely perform duties within hostile, hazardous environments
- Have appropriate equipment
- Have needed resources
- Be healthy enough to perform required duties
- Have mental and internal stamina to withstand catastrophic findings

DISASTER SITE WORKERS—INDIVIDUAL PREPARATION

Not everyone can properly perform under traumatic incident stress situations. When biological hazards, loss of life, and decay are found within a disaster site, workers must be personally prepared from a number of perspectives.

The following checklist contains serious thoughts for disaster site workers to consider:

- Take care of yourself and stay focused upon site hazards for your own safety.

- Take the time to monitor your own physical and emotional health, especially when site recoveries last for extended periods of time.
- Pace yourself, because rescue and recovery at disaster sites may last for long periods of time.
- Mental fatigue that develops over long-lasting work shifts will increase risks for accident and injury—take routine rest breaks.
- The buddy system prevails. Keep an eye open for co-workers that may have lost focus upon overall site activities—they may not be aware of hazards in front or behind them.
- Everyone around you is likely to be tired, stressed and easily distracted. Beware of increased potentials of risk for yourself and others.
- Follow work team schedules, eat, and sleep regularly.
- Drink plenty of water, juices, and other fluids.
- Eat varieties of foods, especially try to increase complex carbohydrates in your diet (breads, grains, granola bars etc…).
- Leave the disaster site whenever possible and try to do normal things in a clean and non-hazardous environment.
- Recognize and accept those things that you cannot change.
- In disaster situations, it's normal to not want to talk to others about your feelings—talk to others when you decide that you are ready. Sometimes talking about a situation may be the same as reliving it. You choose the time and place, if ever, to discuss stressful experiences.
- Sometimes the employer may provide mental health support for disaster site workers, if this service is available - use it.
- It's OK to feel bad about difficult situations; give yourself permission to do so.
- Over time reoccurring thoughts, "flashbacks," will go away, don't try to fight them - let them go.
- Stay close to those that you love and when serving at disaster sites, keep in touch with them as much as you can.

What Happens When Your Work Comes to an End?

Everyone is different in how they adjust back to normal life after serving in a disaster; however there are some basic perspectives that seem to work for most of us. Consider the following:

- Don't be afraid to reach out to others; most people really care.
- Interface as soon as possible with family, friends, the community, spiritual support, and your inner spirit.
- Some folks keep a journal or even write a book.
- Take your time to make any major life decisions.

- Take control of your life; make as many small decisions that you can.
- Recharge your inner batteries, have some fun and spend time doing those things that you enjoy - why do you think so many "old guys" golf so much?
- After witnessing disaster conditions, some people become very fearful for their families. Please know that this is normal. Time will eventually ease these concerns.
- Sometimes returning to our normal selves takes more time than we think it should. In this situation we should learn to delegate our stresses until our shoulders are strong enough to carry them again.
- Laugh again; it's good for the body and the soul.
- Many of us that return home after war, a disaster or other bad experiences have learned the value of support, personal communication, patience, and understanding. Remember, those closest to you have also lived your experiences - consider them.
- Don't do drugs or alcohol. They will only magnify those things that you may be trying to forget.
- Try to get back to normal - what is that? A common sense approach would be to start by: Keep busy, get plenty of exercise, get plenty of rest, eat routinely, and eat stuff that is good for you.

The author sometimes shares a story with his disaster response students about a loving family that had a young son that left home because he wanted to learn about the challenges of the world on his own. Years went by and no one in the family had heard anything from the son. His brokenhearted, aging father saved money and paid to have his son located. When he was found, his father sent a messenger with a letter begging his son to return home to his family. The messenger returned back with a letter from his son that said, "I can never return home. Because of my experiences I have changed and my family would not want me as I have become." The wise father sent another messenger with a letter that said, "Please return back to us as close as you can, and I will come to you the rest of the way." Never discount the love of family.

DISASTER SITE CLEANUP—ABATEMENT OF WATERBORNE BIOLOGICAL HAZARDS

Water is one of the strongest forces of nature. The aftermath of waterborne disaster involves proliferation of microbial hazards that must be controlled for healthy human life to remain.

For a number of years there have been concerns in the United States relating to fungal and bacterial contamination in homes and buildings after floods and other forms of water damage. Unfortunately, the national media and lawyers found financial opportunities in litigating against water damage contractors, home sellers, realty professionals, insurance companies, and others. In this regard, coverage of remediation and decontamination of waterborne biohazards in American homes and buildings have been cancelled by most insurance companies. The problem still exists; biohazard contamination must be controlled for healthy human life to remain.

The following information is presented as considerations and examples of ways that biohazard contamination related to water damage may possibly be handled. The following does not contain any recommendations by the author or any suggested standard operating procedures from anyone. The following are only ideas that disaster site workers may find considerable. The author is not consulting or recommending anything, just passing on information. Along with experience, where did the author obtain this stuff? Take a look at the references listed at the back of this chapter, there have been several organizations involved in clarifying information about biohazards and mold remediation; many are governmental. Also, along with being a mad chemist, the author is a degreed microbiologist - one that is interested in biohazard decontamination.

Site-specific waterborne biohazard abatement plans, procedures, and protocols are usually implemented based upon the extent of water damage and the magnitude of contamination. As conditions at each site may vary, abatement and recovery plans may vary as well.

The following is presented in order to standardize water hazard classifications and microbial contamination. This category format is used to clarify procedures in the contamination abatement plan (CAP):

Water Damage Classifications

This is a classification of water damage that occurs from a clean water source that has been exposed to biodegradable materials for a period less than 24 hours.

TYPE A.1   CLEAN WATER
Type A.1 Water sources do not contain hazardous substances and they may include, but are not limited to:

- Broken water supply lines
- Clean water basin overflows
- Falling rain water
- Melting ice or snow
- Other non-contaminated sources

TYPE A.2   CONTAMINATED WATER
This is a classification of water damage from a source of water that contains quantities of chemicals, microbes, or physical hazards that could cause illness or injury if consumed by humans.

Type A.2 Water can also occur when Type A.1 Water has been exposed to biodegradable materials in excess of 24 hours.

Type A.2 Water includes, but is not limited to:

- Flooding from dishwashers or washing machines
- Toilet overflows that do not contain feces
- Flooding from sump pumps
- Clean water seepage from cracked foundations or slabs (depending upon water source)
- Flooding from water beds or aquariums
- Other similar sources

TYPE B   HAZARDOUS WATER

A water damage source that is unsanitary and contaminated with pathogenic organisms and or hazardous materials is labeled Type B Water.

Type B Water can occur from Type A.2 water that has remained in a spill area for extended periods of time. Lesser hazard types of water that remain in spill areas beyond 48 hours can potentially become Type B water.

Type B Water sources include but are not limited to:

- Water that contains sewage
- Flooding from sea water
- Intrusion by surface waters
- Waters contaminated by hazardous materials, including pesticides, herbicides, toxic organic substances, heavy metals and others

Microbial Contamination Characterization

The following is based upon U.S.EPA Guidelines, and the New York City Health Bureau of Environmental & Occupational Disease, Epidemiology Expert Panel Guidelines on Remediation of Stachybotrys Atra Mold in Indoor Environments.

Abatement procedures and protocols will vary as the magnitude of microbial contamination varies. Layout of containment barriers, worker decontamination corridor placement, levels of abatement activity, and perimeter locations are based upon the extent of microbial contamination. Materials that are contaminated by mold or other potential biohazards are designated as follows:

- LEVEL 1: Small Isolated Contamination = < 10 square feet
- LEVEL 2: Medium Isolated Contamination = < 30 square feet
- LEVEL 3: Large Scale Contamination = < 100 square feet
- LEVEL 4: Extensive Contamination = > 100 square feet
- LEVEL 5: HVAC System Contamination = < 10 square feet
- LEVEL 6: HVAC System Contamination = > 10 square feet

Note: HVAC relates to heating, ventilation, and air conditioning systems. These levels of characterization are mentioned later in this chapter.

Even though abatement professionals may be assigned to evaluate water damage and determine potential microbial contamination in one specific location of a building, the evaluation should be broadened. This is because mold and other microbial contaminants proliferate, especially into obscure locations. In this regard, initial evaluations should be comprehensive. A contamination abatement plan should be developed for each job site. After being ratified, follow the plan. The following should be included when developing contamination abatement plans.

Initial Evaluations

Identify and evaluate indoor and external properties of buildings in the process of determining potentials for biohazards or hazardous materials contamination. Obtain data that can be used to determine potential indoor air quality effects, as well as be used to develop site specific abatement plans. Begin by obtaining information from individuals that are knowledgeable about the building. In some cases it may be necessary to obtain historical data, records, or legal documents about the building and also obtain information about other buildings nearby. Comprehensively evaluate every significant portion of a potentially contaminated building. Note that large contamination areas may require microbiological sampling and analysis protocols to conduct air monitoring before, during, and at completion of abatement procedures. In some cases, the client may request air and surface analysis for all levels of site contamination.

Results obtained from inspection and evaluation activities will dictate activities to be included in the contamination abatement plan (CAP):

- Communicate with occupants
- Build rapport and establish communication with building occupants, building owners, and operators
- Keep communication systems open throughout the entire water damage mitigation, decontamination, mold abatement, problem remediation, and renovation project
- Ask to obtain information
- Employee or tenant air quality-related health issues
- Process or activities performed in the building (past & present)
- Process related hazards
- Hazardous materials used or stored
- Building layout drawings
- Building HVAC drawings
- Previous renovation or abatement activities
- Histories of fire, flooding, or water damage

- Histories of sewer spillage or floor drain backup
- Histories of roof damage or recent repairs
- Any plumbing leakage
- Slab or foundation damage
- Age of building
- Locations that may contain asbestos
- Locations that may contain lead paint or other regulated materials
- Previous mold abatement or problems
- Known or suspect mold or biohazard contamination problems
- Areas that contain unusual odors

FOLLOW SAFE & PRUDENT WORK PRACTICES

- Wear at minimum, OSHA Level D Personal Protective Equipment in contamination Level 1 environments.
- Wear OSHA Level C Total Coverage Personal Protective Equipment in potentially contaminated Levels 2, 3, and 4 environments.
- Do not allow entry into unsafe structures; evaluate structural integrity.
- Inspect using data obtained from initial evaluation findings.
- Do not allow bystanders or building occupants to accompany inspection activities (explain exposure potentials).
- Do not disturb areas of mold or microbial growth.
- Contain any area that contamination may be disturbed during inspection.
- Avoid walking through standing waters.
- Remove hazardous energy, electrical shock, and natural gas hazards from flooded buildings. Follow OSHA Lock Out/Tag out Standards for this procedure.
- Beware of slip and trip hazards.
- Beware of sharp objects.
- Beware of hazards associated with Type A.2 and Type B waters.
- Beware of dangerous animals and insects in obscure locations.
- Practice aseptic, universal precautions: Dispose or contain any clothing or materials that are contaminated during inspection activities. Do not wear or transport any uncontained, contaminated materials away from contamination areas.
- If structures were built prior to 1980, it is important to determine if lead paints are present in areas requiring repair, renovation, abatement, or remediation. Disruption of lead paint requires lead abatement procedures.

- Always investigate potentials for materials that may contain asbestos. If work activities involve disrupting asbestos materials, asbestos abatement procedures will be required to be performed by a licensed contractor.

SITE INSPECTION CHECKLISTS FOR MOISTURE SOURCES, BIOHAZARDS, AND OTHER HAZARDS

- Identify and stabilize uncontrolled potential health and safety hazards before allowing mold abatement activities to resume
- Identify visual mold or other visual microbial growth
- Note unusual or musty odors
- Note indoor plumbing leaks
- Note indoor moisture condensation
- Identify locations of pipes in exterior walls
- Note damp areas on paper-faced gypsum board
- Inspect areas near water heaters
- Inspect areas near clothes washing machines
- Inspect areas near dishwashers
- Inspect areas near clothes dryers
- Look for leakage or evidence of flooding
- Note sewer back up or evidence of sewer back up
- Evaluate bathroom wall tiles, floor tiles, and wall paper
- Shower/bath walls and ceilings for damaged surface coatings
- Evaluate shower and bath walls for plumbing leakage
- Inspect areas under bathrooms
- Inspect for damp basement
- Note damp walls & ceilings
- Inspect crawlspaces
- Look for condensation from pipes and utility surfaces
- Inspect attics for roof leakage
- Evaluate air handling units
- Inspect HVAC System (check lined ducts)
- Look into duct work at access points
- Inspect humidifiers and surrounding area
- Inspect water heater and surrounding area
- Look for construction defects
- Evaluate areas near house plants (due to watering etc.)

- Check kitchens, especially steam areas from cooking
- Inspect concrete and other types of slabs
- Inspect concrete or other types of foundations
- Check for indoor clothes drying lines
- Determine if clothes dryer is improperly vented
- Check to see if combustion appliances are properly vented
- Look for leakage from fish tanks
- Evaluate indoor pet areas
- Check for improper grading of yard
- I.D. flower beds next to exterior walls
- Check for outside sprinklers spraying against the building
- Look for locations of mud or ice dams
- Look for cracked stucco
- ID any clogged weep screeds
- Look for missing or torn moisture paper
- Check for cracked or bubbling paint
- Evaluate stains or discolored surfaces
- Evaluate buckled or warped flooring
- Evaluate distorted baseboards
- Check indoor relative humidity (mold growth >55% RH)
- Look for improperly maintained HVAC Systems
- Check building corners (low insulation level, heat loss)

Oher factors to consider:

1. Dirty and damp air conditioning systems
2. Dirty and damp humidifier or dehumidifier systems
3. Bathrooms without windows or venting
4. Kitchens without windows or venting
5. Refrigerator drip pans
6. Laundry or dryers without venting or vented inside of building
7. Attics with poor ventilation
8. Carpets or rugs placed on damp floors
9. Bedding, mattresses, damp furniture
10. Closets located against a "cold" outside wall.
11. Dirty HVAC system
12. Presence of animals
13. Water leakage from roof, windows, or basement

PERSONAL PROTECTIVE EQUIPMENT FOR SITE WORKERS

The following guidelines are based upon U.S. EPA and OSHA HAZWOPER Standards for protecting workers at work sites that contain chemical and biological hazards:

CLASSIFICATION

OSHA LEVEL D Standard Worker Protection Equipment
Most commonly used for Level 1 Contamination, Water Type A.1, biohazard clean up and remediation activities. May include the following (as directed in specific contamination abatement plans):

- N95 Dust Mask *see note
- Air-purified respirator (HEPA and Organic Vapor Cartridge)
- Rubber gloves
- Eye protection - goggles with sealed fit
- Face shield
- Disposable plastic coveralls
- Standard work clothes
- Hard hat (in designated work areas)
- Splash apron
- Disposable shoe covers
- Disposable arm covers

*Note: Respirator usage must follow OSHA Respiratory Protection Program Standards 29 CFR 1910.134.

OSHA RESPIRATORY PROTECTION STANDARD GUIDELINES

1. Employee must participate in medical questionnaire review
2. May receive a medical checkup (based upon questionnaire)
3. Must be provided a respirator (given a comfortable choice)
4. Must be fit tested (quantitative or qualitative if appropriate)
5. Must receive respirator use & maintenance training
6. Must enter official company respiratory protection program
7. Must be refitted at least annually (sometimes more often, depending upon potentials for hazardous substances that require additional refitting (lead, asbestos & others) and physical changes such as weight gain or weight loss)

8. Must inspect respirators before and after use, and at least monthly if not worn - inspections must be documented and kept

OSHA LEVEL C TOTAL COVERAGE PPE, AIR PURIFIED RESPIRATOR

Used for Contamination Levels 2-4, Water Type A.2 and Type B. Recommended when conducting inspections and evaluation activities. Level C protection shall be used when contamination may exceed permissible exposure limits and unsafe working conditions may exist while performing biological cleanup and remediation activities.
Level C Protection includes:

- Cotton undergarments & socks
- Full face air purified respirator (HEPA, Acid/Organic Filters)
- Boots or heavy duty work shoes
- Plastic, disposable coveralls (with hood & foot coverage)
- Snug fit disposable under-gloves
- Heavy duty plastic, disposable coveralls (with hood & foot coverage)
- Sure grip disposable rubber outer gloves
- Disposable rubber foot covers "Nuke Boots"
- Duct tape (for seal and attachment of gloves & foot covers)

OSHA LEVEL B TOTAL COVERAGE PPE, SUPPLIED AIR RESPIRATOR

Rarely required for biohazard cleanup or remediation. Level B Protection shall be used when suspect hazardous contaminant levels are questionable, or when known contamination is at concentrations immediately dangerous to life and health. Level B Protection is worn when air purified respiratory protection cannot protect employees from chemical or biological hazards.
Level B Protection includes:

- All items listed in Level C Protection except for method of respiratory protection
- Respiratory Protection: Positive Pressure Self Contained Breathing Apparatus (PPSCBA) "Air tanks worn on employees back," or supplied air by hose to respirator including an escape "mini SCBA" respirator also carried by employee.

OSHA LEVEL A TOTAL ENCAPSULATION PERSONAL PROTECTION

Used for extremely hazardous levels of chemicals and biohazards. Often used for cleanup and remediation of treatment, storage, and disposal operations; research operations; and dangerously virulent microbiological environments (terrorist, warfare, WMD etc.)

- Includes donning Level B Protection (no boots)
- Total coverage by a Level A, total encapsulation suit and
- Solid sole, heavy duty, rubber outer boots

The following materials, cleaning products, and equipment are typically used for performing biohazard abatement activities.

MATERIALS & SUPPLIES

- Polyethylene Plastic: Heavy duty (at least 6 mil) thickness for containment; to cover flooring, carpets, furniture for personal decontamination stations
- Duct Tape: for securing PPE, for containment and other uses
- Double Sticky Tape: to attach plastic containment & other uses
- Plastic Glove Bags: at least 6 mil thicknesses, for small area cleanup, sampling, performing tests, evaluations, and other uses
- Containment Booth: expandable and zipper door
- Plastic Disposal Bags: Large, heavy duty (6 mil) thicknesses for disposing of mold-contaminated wastes and damaged materials
- Absorbent: Rolled absorbent padding, pillows, and "pig" booms for absorbing and preventing the migration of wastewaters and chemicals
- Abrasive Pads: Steel wool, Scotch Brite (3M) for loosening and scrubbing contaminants from wood and hard surfaces
- Wipe Cloth: Disposable paper wipes for cleaning mold-affected surfaces
- Warning Tape: Rolled plastic tape used to communicate messages as "Warning, Do Not Enter," "Hazardous Substances" and other information

CLEANERS AND DISINFECTANTS

(Use only EPA authorized biocides & antimicrobials)

- Bleach Disinfectant: Chlorine Bleach is not a cleaner. Contains Sodium hypo chlorite. For maximum strength disinfection, more concentrated mixtures of sodium hypo chlorite solution may be required. A household bleach solution. (.5% chlorine) is commonly used to follow up after cleaning. Common bleach solution mixed with water: mix 1 part of household bleach into 9 parts tap water. Mix bleach and water prior to use, do not mix and store—chlorine content will dissipate and disinfecting ability will be reduced.

Note: bleach mixtures are corrosive; they will irritate the respiratory system and will stain paint and fabric.

Dangerous Reaction: Never use with or around ammonia containing products. (Note: Carefully evaluate before using bleach disinfectants).

- Antimicrobials: Chemicals that limit control or stop the growth of micro-organisms.
- Biocides: A poison that kills living organisms.
- Disinfectants: Chemicals or physical processes that destroy at least 99% of disease causing micro-organisms (Note: remaining 1% may equal 10,000 living microbes/affected million).
- Ultraviolet Lamps: Physical disinfectant process that helps to reduce quantities of microbiological agents.
- Bacteriostat or Fungistat Chemicals: Suppresses the growth of bacteria or fungi.
- Sporicide: A disinfectant that controls or kills microbial spores.
- Sanitizer: A chemical that reduces microbial totals.
- Pesticides: Chemicals that prevent growth or kill pests, including insecticides, rodenticides, herbicides, bactericides, fungicides, and virucides. Please note that governmental agencies regulate the application and use of pesticides. In many cases, the only persons authorized to use pesticide products must be licensed and registered. Use only government registered and approved pesticides, follow MSDS and instructions completely.
- Soaps, Detergents & Deodorizers: Use only high quality products that will not produce hazardous residues or fumes that may cause health problems for building inhabitants.

Note: Closely follow MSDS, technical information and government approvals for all abatement products.

EQUIPMENT AND TOOLS

Water Removal Equipment

- Extraction Units: For removing water from floors, flooring materials, and structural components.
- Pumps: To remove deep standing water.
- Extraction Tools and Attachments: For effectively removing water from flooring and structural components. Typical attachments include light wands, weighted drag wands, vacuum squeegee wands, stair and upholstery tools.

Air Moving Equipment

Used for directing airflow across wet materials and promoting evaporation. Never direct positive air flow at or into potentially contaminated materials; this will spread contamination.

- Air Movers: Squirrel cage fan
- Axial Fans: also called air ejectors, for creating negative or positive pressure within containment or within a building. Sometimes used to direct airflow and in some cases, seal off airflow when appropriate.
- Air Filtration Devices: Portable air movers that contain HEPA or carbon filters. Often used to scrub or filter air while creating negative pressure within containment areas.
- Cavity Drying Equipment: An accessory to air moving equipment. Devices are added to enhance drying by pressurizing or depressurizing structural cavities and causing airflow through leaks in structural materials.

Dehumidifiers

Used for removing evaporated moisture from the air.

- Refrigerant Dehumidifiers: Uses condensation and icing on evaporator coils.
- Desiccant Dehumidifiers: Uses absorption or sticking and is used for a broad range of atmospheric conditions.

Other Drying Equipment

- Supplemental Heaters: For use during cold weather conditions, to raise abatement area temperatures.
- Air Changers: To move indoor and outdoor air through a heat exchanger - useful when outdoor conditions are favorable.
- Vacuum Freeze Drying: To dry out valuable "wet" materials. In this process materials are placed in a freezer and then moisture is sublimed from a frozen state to vapor.

Detection and Monitoring Equipment

- Thermo-hygrometers: Used to determine temperature and relative humidity.
- Moisture Sensors: Uses penetrating probes and by visible or auditory means, determines the presence of moisture in suspected water damaged materials.
- Moisture Meters: Used to determine moisture in specific materials or in combinations of materials.

- Non-Penetrating Moisture Meters: Some use electrical conductivity plates placed on the surface of suspect materials; (+) they don't penetrate the materials; (-) is possible to obtain a false reading.
- Penetrating Moisture Meters: Uses needle probes and measures electrical resistance between the probes (often has additional attachments that add versatility to this device).
- Data Recording Devices: Used to record atmospheric conditions for extended periods of time. May have computer or electronic data collection "hard copy" paper printout, etc.
- Manometer: (Megna-Helix) A digital device that measures static air pressure differential between at least 2 locations of air flow.
- Bore-Scopes: Fiber optics devices that can transfer images to digital cameras to determine magnitudes of hidden mold contamination and water damage. Note: when drilling or boring holes in walls or other materials, consider using a HEPA filtered vacuum simultaneously, during drilling or cutting to prevent spreading potential contamination.

MICROBIOLOGICAL TESTING

If you see or smell mold or microbial growth, you probably don't need to test for it. When cleanup and remediation work is complete, testing may not be conclusive. Why? Because mold and other airborne microbes are everywhere, all the time. In this regard, mold sampling may be similar to taking a picture and capturing a moment in time; immediately after taking samples, the environment is subject to change.

In situations that obvious indoor mold growth is not found, it would be useful to compare indoor air mold spore quantities with outdoor air mold spore quantities. Indoor air should contain less than 80% of total mold spores found outdoors.

In large and dispersed areas of contamination, air sampling should be performed to monitor the efficiency of containment and air filtration activities. Indoor air sampling should also be performed when completing abatement activities.

Along with air quality sampling, the following should be done to determine locations to sample surfaces for fungal growth:

1. Find signs and precursors for mold growth activity.
   - Musty "mildew like" odors
   - Chronic water leaks
   - Damp or dirty insulation in ducts and on pipes
   - Dirty or damp carpet
   - Recent floods or spills
   - Puddles of water near outdoor air intakes
   - Air handling or air conditioning equipment—drip pans

- Indoor soils (plants, basements, etc.)
- Fish tanks, water beds, attics and obscure "damp" areas

2. Look in common areas where mold is found.
   - Basement floors, walls, and ceilings
   - Lower rooms and crawlspaces
   - Any location of water damage or spillage
   - Outside walls and window frames
   - Stained carpet and on carpet backing
   - Ceiling tiles and areas above ceiling tiles
   - Wood structure, wood products, and paper
   - Sheet rock (outside and behind)
   - Under wallpaper and blistering paint

MOLD SAMPLING

Samples are taken for direct examination or for culture testing. When samples are taken, they must be handled using chain of custody protocols from location of sampling to the laboratory. Don't forget! Safe work practices are also required for inspectors.

Typical Sampling Methods Conducted for Direct Examination

- BULK - chunks of suspect mold contamination are carefully placed in a sterile container or a new ziplock plastic bag. This kind of sample is usually tested in the lab for both bacteria and fungi.
- SWABS - sterile swabs are used to collect samples for quantitative purposes. This method of sampling involves rolling a sterile swab across a suspect area and the swab is then rolled over sterile growth media. The petri dish and media are sealed and sent to the lab for analysis.
- WIPES - sterile cloth wipes are used to wipe across suspect materials and then handled for analysis as bulk samples.
- TAPE LIFT - the sticky side of tape is used to touch the surface of a suspect mold growth area and then the tape is pressed against a glass slide. Both the tape and slide are placed into a new plastic ziplock bag and transferred to the lab for analysis.

About Sampling

Microbial testing may include any of the above mentioned methods, including utilizing an open culture plate (called gravity plate) to collect air samples. A problem with using the gravity plate sampling method is that it may likely provide misleading information. Several factors have an effect upon taking air samples:

- Air currents
- Temperature
- Humidity
- Size and weight of airborne spores: larger spores are often overestimated and smaller or lighter spores may not be collected.

Scientific and more quantitative methods involve conducting tests using devices that measure volumes of air as it is sampled. Results obtained from this type of sampling are relative to numbers of organisms per cubic meters of air sampled, sometimes called "sniffer" testing.

Locations for Air Samples

Where should air samples for mold and other micro-organisms be taken? Three locations should be compared:

- Suspect indoor areas
- Outdoors to establish a baseline comparison
- An indoor location, considered O.K. with no complaints

Lab Reports

Microbial sampling and lab reports should answer:

- Who did testing and when?
- Where did samples come from and how were they taken?
- What procedure was used to analyze the samples?
- Results of the analysis?

Microbe analysis reports should not present:

- Any interpretation of results
- Any statement indicating results as good or bad
- Any indication that the mold found is good or bad

Note: Be aware that official regulatory directives for unacceptable levels for mold are being promulgated. To date, no standards "officially" indicate that a site is clean or dirty. When established, the most stringent standards will apply. In this regard, stay in touch with federal, state, and local regulatory agencies.

INTERPRETATION OF MICROBIAL TESTING

The initial assessment of indoor air quality complaints is extremely important. Lab results only fill in certain parts of the puzzle. Along with lab results an effective evaluation should consider:

- General cleanliness of area tested
- Windows or doors normally open or closed?
- Maintenance routine of HVAC systems
- Indoor moisture sources
- Environmental controls for relative humidity
- If people normally occupy areas that were tested
- Due to high levels of activity, sampling may not be reliable
- Normal indoor spore levels range from 30% to 80% of outdoor levels as samples are taken.
- In some cases filtered air, air conditioning, or air from remote outside sources can be as low as 5% of outside spore levels.
- If doors and windows are always open, spore levels may be as high as 95% of outside air levels.
- In dusty areas, indoor spore levels may exceed 100% of outdoor spore levels.

How about the weather?

- Rain can clean mold spores from the air.
- Rain can also assist in dispersing mold spores.
- Samples taken at times that air is high in humidity may not reflect true measures of mold inside as compared to mold sources outside. Spore types and mold species can vary outside during a rainy day verses a dry sunny day.
- Samplings on days that have strong winds also cause analysis interpretation problems. Outside spore counts may be much higher on windy days.

In some cases, outdoor variables mask small to moderate indoor spore readings that may be significant in solving indoor air problems. The reason for air sampling is to compare ratios of indoor mold spore counts with outdoor spore counts. If outdoor conditions are skewed, the indoor vs. outdoor analysis comparison will not be valid.

OTHER SITUATIONS

Because of required sterile environments in hospitals, it is considered appropriate that indoor hospital air contain less than 1 spore per cubic meter of air. Protocols used for testing indoor air in hospitals are much more complex and time consuming than sampling in homes or industrial settings. Most of all, methods and procedures must be correct and analysis must be meaningful.

Sampling in school buildings is an example of problems encountered from high levels of activity. Because of this, microbial tests conducted in schools yield high levels of fungi and bacteria. Financial constraints in recent years have had an effect upon routine maintenance of HVAC systems and general building upkeep. Because of sensitive issues relating to schools, sampling protocols must be exceptionally detailed and thorough.

TAKE A COMMON SENSE APPROACH

Response to water damage and biohazard problems includes:

a. Evaluation of site hazards
b. Following safe work practices, personal protection (PPE) for all entry & abatement personnel
c. Quick response to stop water damage sources
d. Stabilizing water damage and immediately drying the affected area
e. Preventing hazardous exposures to building occupants and mold abatement personnel
f. Identifying the cause of moisture problem(s)
g. Determining the magnitude of chemical or biohazard contamination.
h. Containment of contaminated areas
i. Establishing perimeters and restricting entry into hazard areas; allowing only authorized personnel to enter
j. Installing a decontamination "DECON" corridor for entering and exiting abatement personnel
k. Installing negative pressure, HEPA-filtered air-moving equipment into abatement work areas
l. Disposal of non-salvageable material
m. Cleaning and decontamination of salvageable materials
n. Dry out and disinfection of affected areas
o. Performing microbial sampling when required
p. Remediation, repair, and or replacement of water damage sources

BIOHAZARD ABATEMENT PLANS INCLUDE:

• Meeting with building occupants to discuss water damage and contamination concerns. Keep them appraised about abatement activities through completion of the job.

- Follow safe work practices
- If possible relocate building occupants
- Establish and enforce no-cross perimeters to prevent entry of non-essential personnel
- Establish Decontamination (DECON) zones for entry personnel
- Perform Inspection, Evaluation, and Assessment Procedures
- Perform microbiological pre-work sampling (when required)
- Determine proper abatement methods and procedures
- Determine and use proper personal protection equipment
- Determine and use proper personal protection procedures
- Determine initial abatement equipment needs
- Identify and correct moisture sources
- Remove standing waters
- Determine containment requirements
- Install containment around biohazard removal areas
- Install air-moving equipment, HEPA filters, and control devices
- Bag, remove, and properly dispose non-salvageable materials
- Clean and dry salvageable materials
- Bag and move salvageable materials requiring treatment
- Clean, dry, and disinfect contamination areas (HEPA Vacuum contaminated areas before cleaning and after - including containment, equipment, and PPE)
- Bag, containerize, and dispose all contaminated wastes
- For final decontamination, wipe all abatement areas with disposable disinfectant wipers
- Remove all equipment
- HEPA Vacuum all dusts and final decontamination wipe all areas within and around remediation zones
- If required, perform microbiological sampling of materials in abatement areas and take air quality samples within the building
- Meet with building occupants and provide EPA "How to Prevent Mold Guidance Documents." Have affected inhabitants or responsible parties sign a "read and understood" document that is attached to the mold prevention handouts.
- Conduct a short follow up visit within one week of work completion

CONSIDERATIONS FOR WATER DAMAGE EVALUATORS, INSPECTORS, AND PROJECT MANAGERS

- Evaluate the overall hazards, damage, and magnitude of contamination before committing to loss mitigation and restoration capabilities

- Determine the type of damaging water source (A.1,A.2, B)
- Determine magnitude of potential contamination (Levels 1- 6)
- Have building occupants been exposed to type A.2 or B hazards?
- Are occupants suffering from any health conditions?
- Occupants sensitive to cleaning or disinfecting chemicals, or materials used for abatement or remediation?
- Did water damage occur > 24 hours prior to evaluation?
- Consider Type A.1 water damage after 24 hours, as Type A.2
- Obtain history about property
- Obtain related insurance information
- What steps have already been taken for current water intrusion?
- Nature of affected areas: carpets, furniture, walls, hidden areas

TYPE A.1 CLEAN WATER INTRUSION

If wet < 24 hrs, considerations:
- Abatement personnel wear at least Level D PPE
- Eliminate utility hazards (electrical, natural gas & others)
- Inspect for mold or previous water damage (hidden mold)
- Stop and repair sources of water intrusion (time factor)
- Remove, dry, and secure salvageable items, furniture, drapes, appliances, sensitive items, and valuables (time factor)
- Don't run HVAC (electrical hazards, spread contamination)
- Dispose porous, non-salvageable materials
- Place foil or plastic under legs of remaining furnishings
- Don't run electrical appliances in wet areas
- Don't enter standing water
- Don't place printed or colored paper on wet surfaces
- Make appropriate psychrometry adjustments (temperature & air flow)

TYPE A.2 OR TYPE B CONTAMINATED WATER INTRUSION

Considerations:
- Relocate building occupants if possible
- Discourage occupants that are children, unhealthy, or immune suppressed from re-entering until the building is completely free from contamination.
- Wear at least Level C personal protective equipment

- Turn off utility sources (electrical circuits, natural gas)
- Don't use electrical equipment in wet areas
- Don't use fans to blow dry wet areas
- Turn off and seal HVAC Systems
- Time is critical for stopping water sources.
- Time is critical for water removal and drying water damaged materials.
- Only use HEPA filtered vacuum systems.
- Only use HEPA filtered negative air moving equipment
- Avoid entering standing waters and spillage areas
- Avoid handling contaminated materials
- Establish perimeters and install DECON corridor
- Limit entry to authorized inspectors, evaluators, and decontamination personnel
- Decontaminate all personnel and personal protective equipment when leaving the contamination zone
- Contain, bag, and properly dispose all contaminated wastes and non-salvageable materials
- Dispose all contaminated food and water
- Dispose all contaminated personal hygiene items

LEVELS OF CONTAMINATION

- Any level of contamination should be considered in dealing with Type A.2 or Type B water damage.
- Contamination magnitude, contaminant distribution, and related job site variables constitute prime determining factors for choosing appropriate abatement plans and protocols.

WATER DAMAGE MITIGATION CHECKLISTS

Safety:
- Perform safe work practices
- Conduct hazards assessment and eliminate or control physical and biological hazards as previously noted

General:
- Inventory all affected areas
- Inventory all exposed materials and furnishings

- Note affected carpeting; identify carpets under fixtures, furnishings, and cabinets
- Use a moisture meter to check moisture in water damaged drywall and wood
- Determine relative humidity to all indoor portions of the affected building

Ceiling Tile:
- Remove and dispose tile within 24 hours of becoming wet
- If tile has remained wet for 48 hours or more, evaluate and follow mold abatement procedures.

Drywall or Lathe Plaster:
- Remove all water-damaged drywall and insulation within 24 hours of water damage.
- Use a moisture meter to mark and cut sheet rock at least 12 inches above the moisture mark.
- If there has been previous water damage, or the presence of microbial growth, a mold abatement plan must be established and mold abatement procedures performed.
- If mold abatement is not required, then drying and dehumidification procedures must be implemented.
- Remove and replace all water-damaged drywall and or lathe plaster.

Electrical Devices:
- Electrical power must be shut off in wet areas.
- Circuit breakers, Ground Fault Interrupters, and fuses that have been wet should be replaced.
- Switches, outlets, fixtures, motors and other electrical devices must be opened, inspected, cleaned, and dried. If in doubt, replace them.

Furniture:
- Dispose upholstered furniture that has become contaminated by Type A.2 and Type B waters.
- Upholstered furniture that has been wet by Type A.1 waters may be cleaned and salvaged if treated and dried within 24 hours of water exposure. Monitor for odors and fungal growth.
- Hardwood and laminate wood furniture should be cleaned, dried, and treated with an appropriate and EPA-approved disinfectant. Dispose if delaminated layers separate.
- Furniture composed of fabricated particle or wafer board should be disposed. In some cases, furniture exposed to type A.1 waters that has been immedi-

ately and completely dried may be salvaged if not distorted or swollen from water exposure. If salvaged, monitor for odors and microbial growth.

Paper Goods:
- Dispose non-essential paper goods
- If exposed to type A.1 waters for less than 24 hours, some paper goods may be appropriately dried and salvaged. This often involves moving the water damaged papers to an appropriate area to dry, photocopying and discarding the water damaged articles.
- If essential paper goods have been water damaged and contaminated, they must be handled by technicians wearing appropriate personal protection (see Level C PPE). Proper treatment of contaminated essential paper goods may involve rinsing with clean water and temporarily freezing until proper drying procedures can be performed. Refer to the American Institute of Conservation (202-452-9545) for more information.
- Discard any paper item that develops mold

Carpet: Clean up carpets contaminated with Type A.1 water
- Follow safe work practices
- Pump, absorb, or drain to sewer all standing water
- Inspect for mold in areas affected by the water
- If visible mold or odors are present, do not disturb mold growth. Implement mold abatement plans.
- Remove all items off of carpets
- Extract water from carpet using HEPA Wet Dry Vacuums
- Shampoo with an authorized sanitizing carpet cleaner, following all manufacturers directions
- Steam cleaning can enhance sanitization and drying
- Dry carpet for 12-24 hours after treatment. If possible, increase room temperature, use exhaust fans, and dehumidifiers to reduce relative humidity and dry the carpet and surrounding materials.

WATER DAMAGE CLEAN UP EXAMPLES

Clean Up Type A.1 Water Damage < 24 hrs

- Follow safe work practices
- Abatement personnel should wear at least Level D PPE
- Eliminate utility hazards (electrical, natural gas & others)

- Turn off HVAC system
- Stop incoming water sources
- Don't stand in water
- Remove standing waters
- Don't place or run electrical devices on wet surfaces
- Inspect for mold and other microbial growth from the current occurrence or from previous water damage
- If no microbial growth or previous water damage, HEPA vacuum and use other drying assist equipment to dry the water damage area as soon as possible
- Remove, dry, and secure salvageable items
- Dispose porous, non-salvageable items
- Place foil or plastic under legs of remaining furnishings that remain on carpet
- Don't place printed or colored paper on wet surfaces
- Make appropriate psychrometric adjustments to reduce relative humidity (temperature/air flow etc.)

Clean Up Type A.2 & Type B Water Damage:

- Follow safe work practices
- Abatement professionals shall wear at least Level C PPE
- Eliminate utility hazards (electrical, natural gas & others)
- Contain the contaminated water damage area and isolate from other portions the building
- Keep out of standing waters
- Remove standing waters (pumping or floor drain to sewer) ** Beware: some local ordinances may not allow disposal into sewer; check before disposing**
- Disconnect and seal building HVAC systems
- Exhaust air from contaminated area using HEPA and, if possible, carbon-filtered negative pressure air system
- Aid drying by vacuuming damp materials with wet/dry HEPA Vacuum. Direct negative air exhaust to draw air across wet areas and out of the building (HEPA filter exhaust air).
- Remove and dispose all wet porous materials. Place in sealed heavy duty plastic bags and remove for disposal.
- Remove, clean, and disinfect salvageable materials. (Move to a controlled, contained area to perform cleaning and allow drying).
- Remove carpets, drapes, and other contaminated large items. Place in heavy duty plastic bags or wrap in heavy duty plastic for transporting to waste disposal facilities. (Dispose all contaminated wastes as directed by local waste management authorities).

- Inspect entire water damage area and surrounding areas for microbial growth (mold). Do not disturb mold growth areas until mold abatement plans have been implemented.
- Remove water damaged wall board, unless hidden mold is suspected—include in mold abatement plan
- Wash areas under carpets, large contaminated items, floors, walls, and other areas with a sanitizing cleaner and clean water. Air dry and treat cleaned surfaces with an EPA authorized disinfectant. In some cases bleach and water mixtures may be appropriate, and in other situations, proprietary disinfectants may be warranted. In all cases refer to technical use information provided in manufacturers instructions and pay attention to warnings listed in manufacturers material safety data sheets.
- Dry the area and air if possible, using methods that will not disturb microbial growth, heat lamps etc.
- Don't use positive pressure air or fans
- Implement abatement plans for known or suspected mold or biohazard contamination areas

MOLD & SPORE-FORMING BIOHAZARD CONTAMINATION ABATEMENT PLANS

General Considerations:

- Evaluate occupant complaints
- Evaluate odors
- Evaluate moisture areas, damp materials, remove the moisture, and eliminate a life source for mold and mold spores (use moisture meters, humidity meters, others)
- Inspect all areas including HVAC
- Inspect for visible and hidden mold (bore-scope)
- When mold is found: Evaluate air pressure differentials between microbial growth areas and surrounding areas to determine air pathways and potential impacts upon the overall building and occupants.
- Microbial sampling may not be necessary if visible mold is present. If sampling is conducted, there should be a specific question that test results should answer. Mold testing without a specific purpose may likely produce useless data.
- Some conditions may warrant mold species identification, in this regard sampling is required.
- In some situations related to health issues, litigation, or if sources of contamination are not defined, microbial sampling may be considered as part of the building evaluation.

- If mold is suspected but not visually detected, microbial sampling could help determine potential indoor locations.
- Air sampling for mold spores may indicate that indoor spore quantities are typical or unusual.
- All levels of mold abatement require specific degrees of containment. The lowest level may include sealing contaminants into a plastic disposal bag. Whereas higher levels may include local area or major area containment and isolation - depending upon contamination levels.
- Biocides are not a substitute for thorough cleaning. Biocides kill 99.999% of the mold but leave behind toxic materials produced by the mold. All products of the mold and affected materials need to be removed and disposed. Follow up biocide treatment after cleaning and drying may be prudent. Depending upon the biocide, manufacturer directions for use, and safety and health considerations.

Containment Considerations

- As recommended by the U.S. EPA, always maintain containment areas under negative pressure exhaust.
- Exhaust to outdoors and ensure that adequate make up air is provided. (HEPA filter contaminated exhaust air)
- If the containment system is working, the polyethylene sheeting should billow inwards on all sides. If it flutters or billows outward, negative pressure and proper containment is lost. Correct the problem before continuing abatement activities.
- Minimal containment - Procedures used in performing initial mold inspections, evaluations, sampling and others. Such as cutting small holes in walls to evaluate potentials of hidden mold and others.
- May involve using a HEPA vacuum attached at the cutting site to capture dusts and potential spores.
- Using a glove bag that is sealed onto a surface to conduct sampling or to perform a small operation that may disturb the surface or structure. (For evaluating hidden mold or water damage).
- Limited containment: < 30 Sq ft of contamination
- Single layer of 6 mil, fire-retardant polyplastic
- Enclose area around moldy area
- Slit entry and covering flap (outside)
- Small attachment areas sealed with duct tape
- Larger structures may use framing, etc.
- Seal HVAC components within containment
- Vent using negative pressure (HEPA filtered to outside of building)

- Clean, HEPA vacuum, and disinfect (if appropriate) within the containment area, any items that are to be removed from the containment area
- Begin abatement procedures once negative pressure is constant
- Full containment: (Contamination > 30 square feet)
- For large abatement jobs or any situation that occupant spaces may be further contaminated if full containment is not used
- Double layers of polyethylene plastic should be used to create a barrier between contaminated areas and other parts of the building.
- An airlock or decontamination chamber should be constructed for exiting or entering the abatement area.
- Entry ports from outside entry to the decontamination chamber and from the decontamination chamber to the abatement area should consist of a slit entry with flaps covering the outside surfaces of each slit entry.
- The decontamination chamber should be large enough to hold a waste container, bags for contaminated PPE, a HEPA vacuum, and to allow abatement technicians to put on and remove PPE.
- Respirators should be HEPA vacuumed and not removed until technicians are outside of the decontamination chamber. Take off and place in a sealed plastic bag at the exit point immediately outside of the DECON chamber.

Figure 10.1 Full Containment U.S. EPA.

Note: PPE should be worn in all phases of abatement, including final clean up, including dismantling of containment.

Cleaning Considerations

The following progression is commonly used for room cleaning:

- Ceilings
- Crown, wall molding, and doors
- Picture frames, mirror frames, and headboards
- Hard furniture (desks, nightstands, hutch, TV/VCR etc.)
- Soft furniture (upholstery, bedspreads, curtains etc.)
- Wall surfaces
- Carpets
- Floors

LEVEL 1 CONTAMINATION ABATEMENT PLANS

- Contamination area is < 10 square feet
- Follow safe work practices
- Building inhabitants should not be within abatement area
- Water sources shall be stopped
- Abatement area should be dry
- Abatement technicians should wear OSHA Level D PPE, with dust mask and, in some cases, OSHA Level C total coverage PPE depending upon dispersion of contamination.
- Turn off HVAC system and cover ports if possible
- Special containment measures may not be required, however minimal containment is recommended (see containment considerations).
- HEPA vacuuming should be performed to remove all dusts and particles before and during abatement procedures.
- Remove and dispose porous, non-cleanable contaminated materials. Dispose in sealed plastic bags.
- Scrub contamination with soap and water and dry completely
- Treat cleaned areas with bleach disinfectant in areas that do not contain materials that will react or be damaged by bleach. In sensitive areas use an EPA-authorized disinfectant.
- HEPA vacuum the entire abatement area, including PPE and plastic contamination disposal bags
- Bag and discard disposable PPE

LEVEL 2 CONTAMINATION ABATEMENT PLANS

This procedure includes considerations that are previously listed and additionally includes the following:

- Contamination is > 10 sq ft and < 30 sq ft.
- Follow safe work practices for larger contamination areas
- Use OSHA Level C total coverage (PPE) personal protective equipment at minimum
- Inhabitants should be out of abatement area
- Turn off HVAC system is possible
- Enter only after water damage source has been stopped
- Enter only if no standing water is in abatement area
- Enter only if all internal and external building hazards have been stabilized in or around the scheduled abatement area
- Do not place electrical equipment in wet areas
- Review area for additional visual and hidden mold
- Install limited containment as appropriate, and seal off HVAC system vents and openings with duct tape and heavy duty plastic
- Install filtered negative air exhaust in containment area
- HEPA vacuum all abatement surfaces
- Suppress dusts in areas that require cutting, drilling, or removing by light moisture misting
- Dispose all porous and non-recoverable contaminated items by sealing in heavy duty plastic bags
- Scrub remaining contaminated materials with soap and water; rinse thoroughly and allow drying.
- If mold does not wash away from wallboard or similar materials, cover the contaminated surface by attaching plastic with duct tape, cut it out and replace it. HEPA vacuum dusts during and after cutting.
- For wood: If it is moldy, it will clean by scrubbing with soap and water and allowing it to dry. In some cases proprietary cleaners, disinfectants, and sealers are recommended for wood mold treatment.
- Do not sand moldy wood. Wood that appears moldy and cannot be cleaned is probably rotted.
- Wood rot: Cut it out, dispose rotted wood, and replace with good wood.
- Treated and cleaned wood may be stained. It may be prudent to paint the stained dry wood with latex paint if possible.
- Materials to be disposed: As contaminated materials are collected, place them into heavy duty plastic bags. Seal the disposal bags within the containment area.

- All items removed from containment should be HEPA vacuumed, including outer portions of sealed plastic disposal bags.
- At the completion of abatement procedures, damp wipe all surfaces within containment using one-time-use disposable wiper pads.
- In areas that known or suspect contamination was cleaned, use an EPA-approved disinfectant on dry treated surfaces. Use caution and follow manufacturer's directions, avoid using bleach solution on materials that will react or be affected by bleach.
- Negative pressure exhaust is to remain on until all abatement procedures are completed and materials within containment area are completely dry.
- If microbia,l surface sampling and air sampling are required.
- If possible, make psychrometric adjustments in building air movement, temperature, and humidity control so relative humidity (RH) between 30% and 60%.
- If microbial tests were required keep containment in place until tests have indicated that indoor mold spore counts are significantly less than outdoor mold spore counts, and that surface samples indicate that cleaned surfaces are clear of mold.
- At the end of the abatement job, HEPA vacuum all areas within containment and all surfaces of containment and decontamination chamber materials.

LEVEL 3 CONTAMINATION ABATEMENT PLANS

This procedure includes considerations previously listed and additionally includes the following:

- Contamination area is >30 square ft. & < 100 square ft.
- This contamination area is considered a significantly large abatement area and microbial testing should be performed before proceeding with abatement procedures.
- Microbial testing should be conducted to determine the extent of contamination throughout the building, as well as the effect it may have upon inhabitants. Testing would include: air sampling, surface sampling, and air monitoring during abatement activities.
- Mold abatement technicians should be competent in extensive biohazard and hazardous materials work environments (including trained in HAZWOPER).
- Full containment must be installed to encapsulate contamination areas.
- Full containment should include a full decontamination chamber, airlocks with slit port entry and exit with flaps that cover the outsides of each port.

LEVEL 4 CONTAMINATION ABATEMENT PLANS

This procedure includes considerations previously listed and additionally includes the following:

- Extensive contamination area is > 100 square ft. in one area, widespread areas of control
- More comprehensive decontamination throughout the affected structure
- More control required for all personnel involved in abatement and related activities
- Microbial air sampling and surface testing initially; air monitoring during abatement activities; and air sampling and surface sampling to verify completion of abatement procedures.

LEVEL 5 & LEVEL 6 HVAC CONTAMINATION ABATEMENT GUIDELINES

This procedure focuses upon remediation of mold contamination in heating, ventilation, and air conditioning (HVAC) systems. Level 5 contaminations are less than 10 square feet and Level 6 contamination is greater than 10 square feet.

- Follow safe work practices
- Abatement personnel shall wear total coverage Level C PPE
- Personnel involved in this work should be competent in handling hazardous materials and possess knowledge related to HVAC systems.
- Microbial testing should be performed to indicate microbial reservoirs within the HVAC System. Air sampling should be performed to determine mold spore counts initially, and sampled for monitoring during abatement activities. At the conclusion of abatement activities, surface samples should be taken from access locations within the HVAC system and to compare indoor spore counts with outside levels.
- Visual inspection (also notable odors) should have determined potential contamination points within the HVAC system.
- Full containment, including decontamination chamber with airlocks, is required to isolate HVAC disassembly and abatement areas from other areas of the building.
- When duct work is cleaned, a HEPA-filtered vacuum collector system capable of maintaining up to 1.0 inch of static pressure should be used to vent HVAC system components.
- Only HEPA-filtered wet and dry vacuums are to be used in cleaning contaminated HVAC components.

- Rotary brush and compressed water sprays are commonly used to clean HVAC components. Ensure that these operations are performed within contamination containment areas.
- Isolate abatement areas within HVAC system from occupied spaces by covering openings, fixtures, and other HVAC components with plastic sheeting and duct tape.
- Install HEPA filtered negative air exhaust within containment areas.
- In some cases, cleaning of HVAC duct systems is not practical and HVAC components must be removed. New components are installed after all adjoining components are cleaned and disinfected.
- Contaminated materials removed from HVAC system components must be placed in sealed plastic bags for disposal.
- Large contaminated components removed from HVAC systems should be disinfected and sealed in plastic sheeting prior to transporting through the building.
- Use access openings throughout the contaminated HVAC system in order to facilitate HEPA vacuuming of system components and to perform wet cleaning with authorized soaps and cleaners and to treat with appropriate biocides. See EPA authorized biocides, and recommendations in the New York City Department of Health Mold Remediation Guidelines.
- Clean and dry HVAC system components and coat with authorized antifungal emulsion products.
- Disinfectants may be used after all contaminated HVAC components are cleaned or replaced. Note: follow manufacturer's directions when using disinfectants. Follow recommended times for curing and allowing dissipation of vaporized disinfectant chemistry before activating the HVAC system.

Example Sequence for Cleaning and Decontaminating HVAC Systems

Follow all safety procedures and all considerations listed in previous abatement plans:

- Conduct work during off duty hours in an industrial setting or with inhabitants out of the building
- Interrupt automatic settings for HVAC system and run on manual settings. Do not operate when ducts are being cleaned.
- Clean outside grills and intake ports
- Clean return air ducts, beginning with outer portions of the return air system and stopping at mixing chambers and exhaust stacks
- Clean interiors of air handling unit. Remove and replace interior insulating materials.
- Vacuum diffusers, grills, and registers throughout the ductwork.

- In some cases it may be necessary to remove components, clean, disinfect, and re-install.
- Remaining ductwork should be neatly cut in order to facilitate access to system components. When this is performed, HEPA-filtered negative pressure vacuum collector equipment should be placed downstream to filter dusts and contamination. Side ports downstream should be capped or covered with plastic sheeting and duct tape.
- HEPA vacuum collection should be used throughout all cleaning activities.
- The supply network should be cleaned by starting at the supply fan and ending at the supply diffuser.
- Do not allow unclean duct work upstream from cleaned ducts.
- Cutting additional port locations should be approved by the owner's representative.
- Existing duct access panels shall be used whenever possible.
- After cleaning, ports shall be capped or covered with plastic and duct tape.
- When accessing ducts, visual inspection should be performed before cleaning. In some cases, fiber optic bore scopes may be needed to evaluate obscure areas.
- HEPA vacuum clean, wash, and HEPA vacuum clean again each portion of the HVAC system.
- Lined ducts: the cleaning process must not damage the lining. Often duct interiors are treated with fog-type chemical cleaning treatment and then coated with an approved encapsulant.
- Unlined ducts are HEPA vacuumed, mechanically cleaned, washed, and treated with a chemical fog and dried.

REFERENCES

The following are considered to be reputable guidelines for disaster site work, biological, and mold decontamination:

- *Institute of Inspection, Cleaning and Restoration (IICRC)*
  S100 - Standard and Reference Guide for Professional Carpet Cleaning
  S300 - Standard and Reference Guide for Professional Upholstery Cleaning
  S500 - Standard and Reference Guide for Professional Water Damage Restoration
  S520 - Standard and Reference Guide for Mold Remediation
- *U.S. Occupational Safety and Health Administration (OSHA)*
  Section 5(a)(1-2) of the OSH Act, referred to as the General Duty Clause
  OSHA Standard 29 CFR 1910.94, Ventilation (for General Industry)
  OSHA Standard 29 CFR 1926.57, Ventilation (for Construction)

OSHA Fed Register 59:15968-16039, Indoor Air Quality

OSHA Fed Register 63:1152-1300, Respiratory Protection

OSHA, SHIB03-10-10 A Brief Guide to Mold in the Workplace, October 14, 2003 (13 pages)

OSHA Fact Sheet on Natural Disaster Recovery: Fungi (2 pages)

OSHA, Metalworking Fluids: Safety and Health Best Practices Manual, 68 pages.

- *ANSI, American National Standards Institute*
ANSI Standard 62-2001, Ventilation for Acceptable Indoor Air Quality
- *CDC, Centers for Disease Control and Prevention, NCEH National Center for Environmental Health*

CDC, NCEH Fact Sheet Molds in the Environment, November 30, 2002 (3 pages)

  CDC, NCEH Fact Sheet Questions & Answers on Stachybotyrys Chartarum and other molds, November 30, 2002 (5 pages)

CDC, State of Science on Molds and Human Health, (Testimony to U.S. Congress by Stephen C. Redd, MD, July 2002, 12 pages)

- *EPA, U.S. Environmental Protection Agency*

EPA Fact Sheet, Sources of Indoor Air Pollution - Biological Pollutants, October 18, 2002, (5 pages)

EPA Guidance, Mold Resources, May 20, 2002 (12 pages)

EPA Guidance Publication No. 402-K-02-003, A Brief Guide to Mold, Moisture, and Your Home, Office of Air and Radiation, Publication No. 402-K-02-003, 20 pages.

EPA Guidance Publication, Mold Remediation in Schools and Commercial Buildings, March 2001 (54 pages)

EPA Fact Sheet, EPA Scientists Develop Technology for Detection of Dangerous Molds, May 8, 2001 (2 pages)

EPA Fact Sheet: Flood Cleanup - Avoiding Indoor Air Quality Problems, U.S. Environmental Protection Agency, Office of Radiation and Indoor Air, Publication No. 402-F-93-005, October 2003, (4 pages)

- *AIHA, American Industrial Hygiene Association*
AIHA Fact Sheet, The Facts About Mold, August 28, 2002 (3 pages)
AIHA , The Facts About Mold: A Glossary, June 19, 2002 (4 pages)
- *ACGIH, American Conference of Governmental Industrial Hygienists*
ACGIH Text, Bioaerosols: Assessment and Control, 1999. (322 pages)
- *NIOSH, National Institute of Occupational Safety and Health*
Histoplasmosis: Protecting Workers at Risk. National Institute for Occupational Safety and Health (NIOSH) DHHS Publication No. 97-146, 1997, September, 16 pages.

- <u>Emerging Disease Issues and Fungal Pathogens Associated with HIV Infection</u>, Ampel, N.M. (1996, April-June). Emerging Infectious Diseases, Vol. 2, No. 2. Discusses fungal diseases associated with HIV patients.
- <u>Coccidioidomycosis: A Reemerging Infectious Disease</u>, Kirkland, T.N.; Fierer, J.F. (1996, July-September). Emerging Infectious Diseases, Vol. 2, No. 3.
- <u>Guidelines on Assessment and Remediation of Fungi in Indoor Environments</u>, New York City Department of Health and Mental Hygiene, Bureau of Environmental and Occupational Disease Epidemiology, January 2002, (17 pages)
- <u>Texas DHS, Mold Assessment and Remediation Rules</u>, Texas Department of Health, August 2004; The Texas Mold Assessment and Remediation Rules (25 TAC Sections 295.301-295.338) became effective May 16, 2004.
- Texas DHS, <u>Review of Practices for Mold Remediation</u>, Texas Department of Health. April 2002.

Chapter 11

CHARACTERIZATION OF CBRNE TERRORIST THREATS AND WEAPONS OF MASS DESTRUCTION (WMD)

DISASTER AND TERRORISM

What is a disaster? A disaster could be any emergency that causes human death, damage to needed resources, or damages the environment. A disaster is often an incident that occurs when people expected to respond do not have training or immediate resources required to stabilize emergency conditions. Irregardless if caused from natural occurrences or from the actions of mankind, stabilization of a disaster begins with preparation. Some emergency response personnel believe that disaster lurks in the shadows of those that are unprepared. Terrorism is violence or force used against people or needed resources. Terrorist acts are performed to intimidate and to create fear and disarray in order to draw attention to the terrorist's agenda.

"The superior man, when resting in safety, does not forget that danger may come. When in the state of security he does not forget the possibility of ruin. When all is orderly, he does not forget that disorder may come. Thus his person is not endangered, and his states and all their clans are preserved."

Confucius 551BC - 479BC

CONSIDERATIONS FOR EMERGENCY RESPONDERS AND DISASTER SITE WORKERS

Before this book is published it will be known that the U.S. Department of Labor, OSHA, will have directed that emergency response personnel and disaster site workers shall be appropriately trained and properly equipped before entering disaster sites. In order to do this, OSHA has developed training that is available through authorized OSHA Training Institute (OTI) education providers nationwide. Anyone that is expected to provide emergency service or support at a disaster site should receive this training.

The Federal Emergency Management Agency (FEMA) and Centers for Disease Control (CDC) has also developed training for emergency responders, emergency medical support personnel, and disaster site workers. It is also recommended that anyone expected to provide emergency medical support at disaster sites receive Hospital Emergency Response Training (HERT) for Weapons of Mass Destruction Events. Contact the National Emergency Training Center in Emmetsburg, Maryland and schedule training as conducted by an authorized provider located near your area.

WHAT ARE CBRNE AGENTS?

After America was attacked on September 11, 2001, a number of once uncommonly used terms flooded the media and crept into the vocabularies of many people across

Figure 11.1 When the World Trade Towers were constructed, over 5,000 tons of asbestos was used for lining the structural steel supports. After the disaster occurred, many emergency personnel and disaster site workers did not wear proper respiratory protection while working on or near the site. (Photo courtesy of CCCHST/HMTRI Disaster Site Worker Train the Trainer, Cedar Rapids, Iowa.)

the nations of the earth. In some cases, the nightly news anchors seemed to re-define terms such as "acts of terrorism," "weapons of mass destruction," and others—depending upon how each story was spun. The focus of this chapter is to characterize hazardous properties of substances and conditions that can cause a disaster or may emerge when disasters occur. The author regrets that we must also discuss some of the horrible events that have resulted and are still occurring, as man demonstrates his inhumanity to mankind during this time.

CBRNE agents are substances and energies that have been used and are being used to destroy people and, if not controlled, will destroy our planet. The acronym CBRNE represents:

- Chemicals: (Industrial & Chemical Weapons)
- Biological: (Bacteria, Fungi, Virus & By-Products)
- Radiological: (Radioactive Contamination)
- Nuclear: (Atomic, Neutron, Hydrogen, & other bombs)
- Explosives: (Explosion Reactive and Incendiary Devices)
  - Ease of Access and Availability: Explosives, Industrial Chemicals, Biological Pathogens, and Related Toxins.
  - High Impact and Large Number of Victims: Explosives, Biological Pathogens, Nuclear Explosion, or Release from a Nuclear Storage Facility.

In what form would CBRNE be found?
1. Liquids
2. Gases
3. Vapors
4. Dusts
5. Spores (microbial spore formers)
6. Microbiological and Biotoxin Suspensions
7. Fumes
8. Electromagnetic Radiations, including electricity
9. Solid Articles
10. Malleable, rubber, and plastic-like substances.

How are CBRNE victims exposed?

- By inhaling airborne hazards
- Through eating or ingesting contaminated products
- By absorbing hazardous products through the skin and mucous membranes
- Through open, damaged skin, or through punctures in the skin

What is common to most CBRNE incidents? Panic where people congregate.

Chemical Characterization for CBRNE

When chemical release and contamination occurs and to keep from perishing, the affected population must be able to control and stabilize disaster conditions. After realizing symptoms and without knowing what substances may be involved, response personnel must be able to find ways to detect and quantify chemical contamination. This is often accomplished by using chemical detection methods and procedures such as:

1. Detection papers are often used that will qualify liquids and vapors (similar to pH papers)
2. Colorimetric tubes can be used if the presence of a specific substance is known or suspected
3. HAZCAT Kits are kits that have been prepared for chemical categorization use by the military and for use by a community HAZMAT team.
4. Substance specific test strips that are often assembled into "ticket booklets" (personnel involved in pesticide detection may be familiar with this detection methodology)
5. Electronic metering systems are used onsite to detect airborne, water borne, contaminated soils, and solid objects for chemical hazards.
6. Collecting samples of solids, airborne substances, wipe and swipe testing for laboratory analysis

Sources of Chemical CBRNE may include:

1. Military supplies and surplus materials that may have been used in previous years, several decades before
   a. Chemical warfare agents
   b. Specific examples may include: VX, mustard gas, lewisite
2. Dangerous products used in industrial operations
   a. These products can be readily available, stored in large quantities, and are easy to access via open, out-of-doors storage and within transport vessels. The United States Department of Transportation and Department of Homeland Security have developed training and awareness programs for handlers and shippers of dangerous goods.
   b. Chemicals used in industry are the second greatest threat for use as weapons of mass destruction (WMD).

TICS—COMMON TOXIC INDUSTRIAL CHEMICALS

Chemical CBRNE categorized by disaster potential

1. Nerve agents:

Ammonia [gas & liquid]	Hydrogen Sulfide Gas
Chlorine gas	Formaldehyde Liquid
Hydrogen Fluoride [gas and liquid acid]	Isobutane Gas
Gasoline & Fuels	Pentane Gas
Flammable Solvents	Titanium Tetrachloride
Chlorine dioxide	Phosgene Gas
Propane Gas	Nitric Acid Liquid
Sulfur dioxide gas	Ethane Gas
Hydrogen Chloride [gas & liquid]	Sulfuric Acid (Oleum)
Hydrogen Gas	Ethylene Liquid
Methane Gas	Vinyl Chloride
Butane Gas	Tri-chlorosilane
Ethylene Oxide	Methyl Chloride

Figure 11.2 Typical industrial chemicals that are shipped and stored in bulk quantities, often in open view, and subject to the environment. USEPA: Accidental Risk Data.

- These products act upon all elements of the nervous systems of people, other animals, insects, and plants. As noted in pesticide use, a nerve agent interrupts the normal transport of electrical nerve impulses within a living system and may cause death.
- Common chemical nerve agents include: VX; sarin; soman
- Symptoms: People involved in a chemical disaster need to be able to identify common warning signs indicating the presence of nerve agents. Unfortunately one warning sign is the obvious effect as the onset of symptoms in victims are manifested.

What are common symptoms of people that have been exposed to hazardous doses of nerve agents?

- Salivation, uncontrolled and choking the victim similar to a person suffering from rabies
- Lacrimation, watery eyes, and uncontrolled tears
- Urination, abnormal rate of urination as the body tries to rid the toxin
- Defecation, uncontrolled diarrhea, and rectal pressure
- Gastrointestinal, severe stomach pain, cramping, and nausea
- Emesis, gagging, and vomiting
- Miosis, pupils of the eye become pinpoint in size

The Release of Nerve Agents

As mentioned above, warning signs for a release of nerve agents upon a population can be easy to observe if people know what to look for. Nerve agents can be drastically lethal, and usually mass fatalities occur without other signs of trauma. Other warning indicators of nerve agent releases may include:

- Chemicals or chemical-containing equipment may be found in areas that are not common for them to be located. Spray containers or related equipment may be found abandoned.
- People may complain of unusual environmental conditions such as odor, tastes, ringing ears, and other sensations.
- An explosion may have occurred that appears to have dispersed powders, liquids, gas, vapors, or mists into the environment.
- A minor explosion may have occurred that only ruptured a package or a containing device.
- Noting the presence of dead animals, insects, and plant life
- Human casualties without obvious trauma, which has common symptoms and demonstrates a distinct pattern

Blister Agents

- These chemicals cause eye irritation, discoloration of affected skin, blisters, respiratory system failure, and depression of the central nervous system and maladies of the gastrointestinal system.
- Common blister agents include: mustard gas, lewisite, and phosgene oxime.

Warning signs that would indicate that blister agents are involved in a disaster may not be as obvious as those mentioned about nerve agents. Often symptoms and blisters occur several hours after a victim has been exposed. Sometimes victims may report eye and respiratory irritation along with noting garlic-like tastes or odors. As would be expected, several victims may begin reporting similar symptoms as well as observations that may indicate blister agent release.

What could indicate a blister agent release?

- A package, ruptured container, or bomb-like device that released powders, liquids, vapors, mists, or gas
- Unscheduled spraying of an area, especially with aircraft "Crop Duster" and similar equipment
- Abandoned spray-related equipment or devices
- Widespread or mass casualty involvement without trauma or other causes
- Symptoms and patterns of related casualties that are common and have a distinct pattern of occurrence

Blood Agents

- These chemicals interrupt the oxygen transfer chain from the blood to tissues. In this regard an affected living system would suffer symptoms similar to anoxia.
- Common blood agents would include : pesticides (furadan), cyanogen chloride, and hydrogen cyanide

Warning Signs for the Presence of Blood Agents
One obvious but usually unnoticed warning sign for the presence of blood agents is a bitter or nutty odor; especially noted in cyanide releases is the initial odor of burnt almonds.

Choking - Pulmonary Agents

- These chemicals cause eye irritation and respiratory failure from swelling of the tissues in the lungs, also called pulmonary edema and chemical pneumonia.
- Some common examples of these chemicals include: gases and vapors of ammonia, chlorine, and phosgene.

Warning Signs for the Presence of Choking Agents
Initial warning signs may include burning and tearing in the eyes and tightness in the chest with belabored breathing. People may report initial odors similar to that of chlorine bleach and, if the release is from phosgene, an initial odor may be similar to that of freshly mowed grass. If the release is a choking agent, the initial odor rapidly changes and symptoms become severe.

Irritants—Substances Used in Riot Control

- The function of these substances, as liquids, gases or vapors, would be to incapacitate the victims rather than kill them
- The effect upon humans by these substances include eye irritation and tearing, skin discoloration, burning sensations in the eyes and mucosal tissues; also victims may suffer from chest tightness, coughing and vomiting.
- Some common examples of chemical irritants include: mace, tear gas, and pepper spray.

Warning Signs for the Presence of Irritating Agents

Exposed people will complain of burning, itching eyes, or irritation in the throat. If victims come in contact with powders, liquids, vapors, or mists, the skin may produce a burning sensation; the lungs will burn and upper gastrointestinal pain may be induced. Exposure to irritants usually involves tearing, coughing, choking, difficulty in breathing, nausea, and vomiting. Other warning signs include:

- A noted peppery odor in the air or on the clothing of exposed victims
- Reports of odors that accompany tear gas, also casings or munitions used to disseminate tear gas may be present.
- Brownish-colored stained materials, structure, or concrete
- Other associated odors such as solvent similar to hair spray or paint propellant, apple blossoms (sometimes used in mace), and pepper odor

When considering potentials of chemical-related materials that may be used for terrorist actions or for use upon large populations as weapons of mass destruction (WMD), a key concern is how lethal a given substance may be. In the following table, a comparison of lethality is shown between various chemical agents and the toxic gas, Chlorine.

Note: The immediate danger to life and health (IDLH) concentration of chlorine gas is 10 ppm. References: NIOSH Pocket Guide to Chemical Hazards and the International Chemical Safety Guides.

BIOLOGICAL CHARACTERIZATION FOR CBRNE

Disasters involving microbial agents may not be manifested as abruptly as chemical agents. One message that is contained in the FEMA Hospital Emergency Response Training for Weapons of Mass Destruction is for area and regional health officials to monitor disease trends within their respective areas and to report suspect incidents immediately.

Biological disease patterns may evolve through:

Table 11.1 Comparison of Lethal Airborne Chemicals and Chlorine Gas.

1. Cyanogen Chloride: 2 times more fatal than chlorine gas
2. Phosgene: 6 times more fatal than chlorine gas
3. Hydrogen Cyanide: 7 times more fatal than chlorine gas
4. Mustard Gas: 13 times more fatal than chlorine gas
5. Sarin: 200 times more fatal than chlorine gas
6. VX: 600 times more fatal than chlorine gas

- A population that is exposed to a toxin that was produced by a microbial agent such as mycotoxins produced by pathogenic fungi, or bacteriotoxins produced by pathogenic bacteria (botulism poisoning).
- Disease etiology or microbial infection that is being spread throughout a population, especially one of questionable origins (smallpox in the U.S). In this regard, symptoms in victims may not occur for extended periods of time and tracking the disease may be difficult. In today's world, one that is blessed with mobility, an epidemic may become pandemic in just a few days.

Significant Biological Disaster—Is It Imminent?

- Terrorist attackers would warn the victims? No way, they usually make efforts to boast and take credit in the aftermath of a terrorist event.
- Health officials would report disease trends and a non-routine number of victims having similar symptoms.
- It would be reported that unnatural disease patterns and atypical distributions of victims are being seen within a given area.
- In some cases, high incidents of disease found in animals such as birds or rats should be reported—West Nile Virus, Black Plague and others.
- Increased disease incidents would be noted in geographic regions not normally seen.
- It may be noted that airborne or unscheduled sprays of liquids or dustings of powders have occurred within a population area, especially if spraying was performed during evening darkness.
- Abandoned equipment or containers may be found, such as biological laboratory supplies, culturing materials, and transfer containers.

How Are Biological Agents Detected, Confirmed, and Quantified?

People that initially respond to evaluate potential biological agents do not have many resources that yield reliable quantitative field test results. Remember the "on-the-spot" anthrax testing that was used in the aftermath of September 11, 2001, which provided unreliable test results? For the most part, initial response personnel are required to collect samples of potential biohazards (microbial and chemical by-products) and send the samples to laboratories in order to conduct controlled chemical analysis and biological assay tests on samples of suspect agents.

When victims receive medical care, the physicians and practitioners will take samples for laboratory testing and treat symptoms until more information is received. What's the big concern? Most hospitals or community support agencies do not have the equipment or bed space to handle a community that has become victim to a major epidemic or biological terrorism.

U.S. Centers for Disease Control and Prevention - Characterize Potential Biological Hazard Agents

The United States public health system and primary healthcare providers are required to be prepared for varieties of biological agents; especially microbial pathogens that are rarely seen in the United States. The highest priority is placed upon pathogenic agents or their bi-products that may risk national security.

Category A Biohazards

- These biohazards can be easily disseminated or transmitted from person to person.
- They can produce high rates of mortality and can exert a major public health impact for care, resources, and support.
- Category "A" Agents have the ability to cause public panic, social disruption, and require special actions for public health preparedness and planning.

Category B Biohazards

- These agents are moderately easy to spread throughout a population.
- They may result in moderate rates of morbidity and low rates of mortality.
- Category "B" Biohazards requires CDC to enhance normal diagnostic capabilities and requires elevated monitoring and disease surveillance.

Category C Biohazards

- These agents have the ability to be biologically engineered to increase potential pathogenic properties and to be easily distributed throughout a population.
- They are readily available and easy to obtain.
- These pathogenic microbes are easy to produce.
- They have high potentials for morbidity and mortality and could produce major health impacts.

CHARACTERIZATION OF CHEMICAL & BIOLOGICAL (CBRNE) AGENTS

If a disaster involves Chemical Nerve Agents the following occurs:

- Increased numbers of dead animals, insects, and wildlife
- Increase of unexplained human casualties having symptoms of serious illness ranging from nervous system disorders through general malaise
- Human casualties will begin to form a pattern of occurrence.
- The presence of unusual powders, liquids, sprays, or vapors may be noted such as: droplets or oily films within the environment, unexplained odors, or the occurrence of high vapor density clouds or fogs that are not related to a weather pattern.
- Findings of suspicious packages, abandoned spray devices, or unexplained munitions

Note: Table 12.2–12.5 list characterization criteria for typical chemical nerve agents that could be used as weapons of mass destruction.

Toxic Industrial Chemicals (TICS)

If TICS are involved in a disaster, the following should be considered:

1. As mentioned earlier in this chapter many chemicals have been created for using in industrial processes. The most dangerous of the "TIC" substances are those that can yield destruction and harmful human health conditions in short periods of time.

2. Victims and emergency response personnel that deal with concentrates of "TICS" as well as those that may be located downwind of a volatile substance, have high potential for serious injury. When large quantities of dangerous substances are released, people closest to the release hold the greatest potential to suffer extremes of hazardous effects. Question: How large would a dangerous quantity of hydrogen cyanide gas need to be if it is released near people in a closed room? Answer: Not very much.

3. Acts of sabotage that would emit TIC Agents from large containers such as storage tanks, train cars, tanker trucks, especially if the TIC entered the air, could yield thousands of casualties. You can't see a need for a Department of Homeland Security? The author warns "look again while you can still see." TIC transportation and storage would be considered a weak target for a prepared terrorist, irregardless of the terrorist's origin.

The following figures are examples and considerations for toxic industrial chemicals, "TICS" if used as weapons of mass destruction.

Table 11.2 The following figures characterize chemical blister forming agents that have been used for weapons of mass destruction, some used before World War I (1918). Mustard Aerosol and Liquids, often called mustard gas; the arsenical chemical Lewisite (L) and Phosgene oxime (CX).

Types of Chemical Nerve Agents	Tabun (Military Agent GA), Sarin (Military Agent GB), Soman (Military Agent GD), Military Agent GF, and Military Agent VX. These are chemical agents that attack the nervous system by inhibiting nerve impulse transfer chemistry (cholinesterase). After exposure, the serious health affects upon the victim may become irreversible. Nerve agents may enter the human body by absorption through the skin, inhalation into the respiratory system, ingestion by eating and sinus drainage, and through openings in the skin or skin punctures.
Victim Response & Symptoms	All Modes of Exposure: Miosis (pinpoint eye pupils), nausea, vomiting, physical weakness, loss of conscious ness, convulsion, flaccid paralysis (limp muscles), secretions of body fluids from most sources, eventual heart, and respiratory failure.
	Aerosol Exposures: Along with the above symptoms aerosol exposures would be indicated by watery eyes, nose, and breathing difficulties.
	Skin and Tissue Exposures: Along with those listed above, symptoms would also include irritation or burning of the skin and localized sweating.
Chemical Characteristics	• Clear to light brown liquids
	• Volatile to non-volatile (wanting to become air at room temperature)
	• Tasteless and no odor to mild fruit like smell.
	• G Agents have a tendency to break down into the environment and may be less persistent, unless prepared as heavy, syrup like mixture.
	• V Agents are less affected by the environment and are reported to be more persistent.
	Note: Alkalis, as well as chlorinating compounds have been used for decontaminating equipment. Dilute chlorinating compounds, soap, and water cleaning are usually used to decontaminate victims.
Treatment	Usually accomplished by decontaminating victims, first aid, and treating with substances such as atropine, pralidoxime chloride, diazepam and others.

Table 11.3   Characterization of Mustard (HD).

MUSTARD (HD)	
Description	Substances that contain Mustard (HD) will burn human skin and other tissues. They damage the eyes, mucosal tissues, lungs, and respiratory system; mustard chemicals may also affect organs that take part in the formation of blood cells.
Victim Response & Symptoms	Victims of mustard chemical exposure will suffer extreme respiratory irritation, watery and burning eyes, bleeding from sinus and nose, coughing, loss of voice tone, reddening and burning of skin. As exposure time increases, the eyelids will swell, coughing becomes painful, and blisters form on exposed tissues.
Physical Characteristics	Mustard chemistry is an oily liquid. It may appear clear to dark brown - depending upon how it was processed. The specific gravity of mustard solution is heavier than water; however, finely dispersed droplets or aerosols can float on water as well as contaminate most materials. The odor of mustard chemistry is similar to garlic or newly ground horseradish. The potential damage from mustard chemistry depends upon the method it is dispersed, how concentrated the contamination was spread, and the temperature of the environment where it was released. Mustard contamination remains longer in absorbent environments, such as wooded areas, and is more volatile and damaging in warm climates than cool climates. However, mustard contamination will remain longer on materials in cold weather.
Modes of Exposure	All modes of exposure are affected. Mustard chemistry will penetrate the skin and all other exposed tissues. Damage will occur directly to exposed tissues and can become systemic. In some cases, immune system failure and related infections are key contributors in causing death to exposed victims.
Basic Treatment of Victims	Proper decontamination requires large quantities of water. Decontaminated skin is usually treated similar to burn victims. In some cases calamine lotions may be used. Approved ointments are used on eyelids, lips, and other damaged tissues. Antibiotics and antibiotic skin treatments are used to prevent the spread of infections.

Table 11.4   Characterization of Lewisite (L).

LEWISITE (L)	
Description	Lewisite (L), a blister-causing agent, is an amber to dark brown liquid. It readily dissolves in most organic solvents and very slowly dissolves in water. As compared to Mustard chemicals, Lewisite will volatilize or become part of the air quality faster than Mustard chemicals under similar conditions.
Victims Response & Symptoms	Symptoms occur immediately following exposure. • Skin: pain and irritation begins within seconds, redness occurs within 15 to 30 minutes. Blisters occur within several hours. Note: Usually heal faster than blisters formed by mustards. • Eyes: irritation, pain, swelling, and watery eyes occur immediately. • Respiratory tract: mucous formation, watery discharge, sneezing, hoarseness, nose bleeds, sinus pain, shortness of breath, and cough. • Digestive tract: diarrhea, nausea, and vomiting. • Cardiovascular: Low blood pressure and shock may occur.
Physical Characteristics	Lewisite will quickly form arsenic-oxides when it reacts with water. • The vapor density is heavier than air and settles in low areas. • Reacting as a liquid, it will penetrate most impermeable fabrics and rubber. • Has a geranium-like odor. • It contains arsenic, a toxic chemical. • The military designation for Lewisite is "L." • Lewisite will remain as liquid below freezing to very hot temperatures. In this regard, it will remain for long periods in the environment.
Modes of Exposure	Exposure to Lewisite occurs in all modes of exposure
Basic Treatment of Victims	• Leave the area and move to fresh air. • Remove exposed clothing and quickly wash entire body with soap and water. • Avoid handling of contaminated clothing without personal protective equipment. • If eyes are exposed and vision is blurred, rinse eyes with plain water for 10 to 15 minutes. If victims wear contacts, do not place back into eyes. • Medical personnel may choose to treat victims with British anti-lewisite (BAL) (dimercaprol) or other antidotes for Lewisite poisoning.

Table 11.5 Characterization of Phosgene oxime.

PHOSGENE OXIME (CX)	
Description	Phosgene oxime (CX), also called a vesicant or blister-causing agent. Phosgene oxime (chemical name dichloroformoxime) is a class of chemical agents called urticants - also called nettle gases.
Victims Response & Symptoms	Initial symptoms: Immediate pain to exposed tissues. • Pulmonary and eye burning, rhinorrhea and coughing. • Skin erythema • If untreated, conjunctival damage and pulmonary edema • Swollen portions of the skin and skin necrosis.
Physical Characteristics	Phosgene oxime exhibits an overpowering penetrating odor. • Can appear as a clear liquid • Colorless, low temperature melting, water soluble crystal. • Compared to mustard chemistry, it immediately produces pain.
Modes of Exposure	Phosgene oxime can expose victims by all modes of exposure. Most commonly it causes pain and irritation to the skin and mucous membranes.
Basic Treatment of Victims	Immediate decontamination, flooding with large amounts of water. Treat symptoms as they appear. Blisters are usually treated with soothing and antibiotic lotions. Small blisters are usually left intact; however, large blisters may require drainage. Treatment for pain, monitor and treat infection and systemic disorders as required. Treat damage to eyes and surrounding tissues. Pneumonitis may occur and may require supportive care, such as oxygen and intubation.

Chlorine is characterized as a TIC hazard as well as a hazardous by-product that can be released by reactions. Question: What occurs when chlorine bleach comes in contact with a cleaner that contains ammonia? Answer: Lots of chlorine gas. Another question: What happens when a train car filled with phosphorus tri-chloride or phosphorus oxy-chloride is release into humid air? Answer: The formation of large amounts of toxic, corrosive gases.

Biological Hazards

Potential disaster threats include:

1. Having the ability to recognize a biological hazard threat.
2. Can a credible threat be identified?
3. The presence of biological terrorist evidence such as devices, or the presence of a biological agent or presence of bio-toxic products.
4. Credible diagnosis of disease, a collection of public health data, or laboratory analysis to verify a potential bioterrorism agent.
5. There will be a delay between victim exposure and manifestation of disease symptoms, an incubation period. Exposure victims may travel great distances during the incubation period, adding to a concern of exposing hundreds of new victims.
6. Biological terrorism agents can also be released for the purpose of affecting food and agricultural commodities that a population may rely upon.
7. Investigation unusual occurrences of dead or dying animals.
8. Investigation of unusual disease occurrences such as: Illness normally no found in a region, or a unnatural pattern of disease etiology exists in a given location.
9. Investigate the presence of unusual liquids, sprays, or vapors.

The following figures are examples and considerations for the use of biological hazards as weapons of mass destruction.

CHARACTERIZATION OF RADIOLOGICAL, NUCLEAR, AND EXPLOSIVE CBRNE AGENTS

How likely will weapons of mass destruction (WMD) be used on people in free world nations? Unfortunately, all people in the world today must develop contingencies to deal with potential threats, especially those of us that fortunately live in free world nations. The impact of a WMD release will depend upon the properties of the CBRNE agent that is used, as well as the methods of deployment. Thus far, we have reviewed and characterized some common chemical and biological WMD agents.

Table 11.6 Characterization of Chlorine as a Toxic Industrial Chemical.

CHLORINE GAS & COMPOUNDS

Description	Chlorine is a greenish-yellow colored gas that has an overbearing pungent odor. Chlorine or chlorine containing compounds can react with other substances to produce toxic and corrosive hazards. Many combustible substances will react with chlorine to cause fire or explosion. Harmful concentrations of chlorine are rapidly reached as it is released from containment.
Victim Response & Symptoms	At 10 ppm, chlorine is immediately dangerous to life and health, IDLH. Chlorine and chlorine compounds are severe inhalation hazards to people and animals. Chlorine is corrosive to all human tissues and inhalation will cause pneumonitis and pulmonary edema (swelling and liquid formation in lungs). High concentrations on human tissues will result in deep burn tissue damage. If victims are not immediately moved to fresh air, a condition called RADS, Reactive Airways Dysfunction, will occur and victims will die. These effects can be delayed and occur hours after exposure. Chronic effects are also bronchitis and erosion of the teeth.
Physical Characteristics	Chlorine gas is heavier than air and will migrate to low lying areas within a surrounding release area. In mixtures with water, chlorine will react violently with basic substances and is corrosive. As mentioned above, it will react with combustible chemicals to cause fire and explosion. Chlorine will attack certain plastics, rubber, normal protective coatings, and metal. Aerosol chlorine or chlorine compounds will combine with moisture to form hydrogen chloride (hydrochloric acid HCl).
Modes of Exposure	The primary modes of exposure to chlorine are inhalation and absorption through the skin and mucosal tissues.
Basic Treatment of Victims	Immediately move victims to fresh air. Position in a half-upright position. Provide artificial respiration as chlorine has likely been inhaled. Rinse contaminated skin with large amounts of water, remove all contaminated clothing and personal articles, and rinse again with large amounts of water. Remove contact lenses. Treat all exposed tissues for burn damage.

Table 11.7   Characterization of Cyanide as a Toxic Industrial Chemical.

CYANIDE (HCN & Compounds)

Description	Cyanide and cyanide-generating substances are toxic blood agents. When hydrogen cyanide enters the circulatory system it interferes with cellular oxygen transfer. It may be said that one of the contributors to the death of cyanide victims is chemical-induced anoxia and eventual failure of all biological systems.
Victim Response & Symptoms	Without immediate treatment, victims of cyanide poisoning can die within 2 to 4 minutes of exposure to significant doses of cyanide. The amount that is immediately dangerous to life and health is 25 mg/cubic meter. After exposure, the victim becomes dizzy, nauseated, and respiratory rates decrease. The victims then experience convulsions, cessation of breathing, and cardiac arrhythmia. Overt symptoms include: • Burning and dryness in the throat. • Shortness of breath or dyspnea. • Rapid shallow breathing or hyperpnea • Lack of breathing or apnea • Convulsion and coma • Cardiovascular collapse.
Physical Characteristics	Cyanide compounds are often white, hygroscopic crystalline powders that have a distinct odor. Hydrogen cyanide is a colorless, highly volatile liquid and hydrogen cyanide gas is liberated when acids react with compounds such as sodium and potassium cyanide. Hydrogen cyanide has a faint odor, similar to peach kernels or bitter almonds, and sometimes cannot be detected even in lethal concentrations. Because the liquid is highly volatile, in the aerosol, gaseous form it dissipates quickly into the environment. As already mentioned, cyanide compounds rapidly decompose when in contact with acids and slowly when in contact with moisture or carbon dioxide. Water solutions of sodium or potassium cyanide are medium to strong bases.
Modes of Exposure	Cyanide exposure can result in all modes. Hydrogen cyanide gas is rapidly absorbed through inhalation, ingestion, through the skin and other tissues. If high concentrations are inhaled, only a few breaths may be enough to cause death. Victim's skin color may become pink and fingernails become bluish "cyanotic" color.
Basic Treatment of Victims	Decontaminate with large amounts of water if skin or clothing has become contaminated by non-gas forms of cyanide. Victims are usually treated by intravenous sodium nitrate and sodium thiosulfate and respiratory support by administering oxygen.

Table 11.8   Characterization of Phosgene as a toxic industrial chemical.

PHOSGENE GAS	
Description	Phosgene is a clear to yellowish-colored, heavier than air, toxic, and corrosive gas that rapidly damages the respiratory systems in humans and animals. Phosgene causes pulmonary edema that produces the sensation of choking and eventual death. The immediately dangerous to life and health (IDLH) quantity of phosgene is 2 parts per million (ppm).
Victim Response & Symptoms	Initial symptoms: • Eye irritation, burning and damage to the conjunctiva. • Pulmonary edema, rhinorrhea, cough, dyspnea, and tightness in the chest. • Swollen patches on the skin, and necrosis of exposed tissues. • Choking and death.
Physical Characteristics	• Phosgene has an odor that resembles freshly cut grass. It will dissolve in organic solvents, fatty oils, and water. When mixed with water, it will form hydrochloric acid and carbon dioxide. • Because it is heavier than air, phosgene can crawl across the ground, remaining for long periods of time in low-lying areas. • Phosgene will decompose at temperatures greater than 300 degrees C and produce toxic and corrosive hydrogen chloride and carbon monoxide gas and chlorine-containing vapors. • It reacts violently with strong oxidizing chemicals, and slowly with water to produce corrosive and toxic gases. • Phosgene reacts violently with amines and aluminum and, in the presence of moisture, will attack metals, plastic, and rubber. • Harmful air concentrations quickly occur when phosgene is released to the environment.
Modes of Exposure	• Phosgene primarily enters the body by inhalation and, in some cases, through absorption. • When inhaled, phosgene travels through the respiratory system. Respiratory exposure damages respiratory system components, especially the alveoli in the lungs that controls oxygen and carbon dioxide exchange for all cells in the body. • Long term, low level exposure can cause respiratory fibrosis.
Basic Treatment of Victims	Move victims to fresh air and keep victims calm and at rest. Treat for respiratory illness. Caution: if victims do not remain at rest until receiving treatment, the harmful effects of phosgene can be increased by exertion and the victim may die. If non-gaseous corrosives or toxins on clothing or the body, remove clothing and decontaminate using large amounts of water.

Table 11.9 Characterization of Anthrax as a Biological Weapon of Mass Destruction.

ANTHRAX	
Description	Anthrax is a highly lethal infection caused by gram-positive bacterium infection, from Bacillus Anthracis. In natural disease etiology, the spore-forming organisms usually gain entrance through openings in the skin. However, anthrax exposure can also occur by inhalation and ingestion. Terrorist activities may also utilize aerosol methods of disseminating anthrax spores. Anthrax spores can remain viable in the environment for long periods of time. The incubation period for disease from inhalation exposure to anthrax is 1-6 days.
Disease Symptoms	• Fever, malaise, fatigue, cough, and mild chest discomfort. • Rapid proliferation of severe respiratory distress, accompanied by dyspnea, diaphoresis, croup-like cough, and cyanosis. • Shock and death may occur within 24-36 hours from the onset of respiratory and other related symptoms. • Skin anthrax infections usually produce swelling and lesions. Without treatment the disease will progress to cause septicemia, which can be fatal in 20% of infected victims. • In treated cases of skin anthrax, fatalities are rare.
Preventive Treatment	A vaccine is available for persons having risk of exposure. If an individual has been exposed, antibiotic therapy is also used - for example doxycycline and ciprofloxacin are commonly used.
Treatment Considerations	Exposed victims are treated by antibiotics and provided needed support. Also, in cases involving skin infection or where drainage or secretions occur, contaminated materials should be collected and decontaminated. Instruments, other equipment, and materials should be decontaminated using a sporicidal agent such as a strong disinfectant like iodine or .5% sodium hypochlorite solution.

Table 11.10   Characterization of Botulism Toxin as a Biological Weapon of Mass Destruction.

BOTULISM	
Description	Clostridium botulinum, botulism toxins consist of seven neurotoxins defined as types A through G. They are produced by the anaerobic bacteria, Clostridium botulinum. For hundreds of years people have died from improperly canning low acid vegetables and meats which contained botulism spores. When canned, the spores become active and, as the bacteria grow within the foods, botulism toxins are produced. Botulism toxins can be destroyed by subjecting to boiling temperatures for extended periods of time. Botulism toxins are extremely toxic. • Terrorists could produce aerosol mixtures of the toxin and release it to victims that would be exposed by inhalation. • Food and water supplies could also be contaminated and victims would be exposed by ingestion of the toxin. If large numbers of victims are subjected to relatively low concentrations of botulism toxins, many casualties will evolve. Because most medical treatment facilities do not have access to large amounts of botulism anti-toxins, a terrorist action using this agent could result in a significantly lethal event.
Disease Symptoms	Botulism toxins react to block acetylcholine in the nervous system. Symptoms of exposure would include: • General weakness with blurred vision • Dry mouth and throat • Flaccid paralysis (extremely limp muscles) • Progressive weakness, respiratory, and heart failure • Symptoms begin within 12 to 36 hours after exposure
Preventive Treatment	Preventive anti-toxins are available for toxin types A through E. Persons that have potentials for exposure may receive preventative doses over a period of 12 weeks and may be given annual boosters.
Treatment Considerations	The presence of Clostridium Botulism micro-organisms can be decontaminated using .5% hypochlorite bleach disinfectant. The toxin can be cleaned from surfaces using detergent solutions; however contaminated cleaning solutions require destructive treament of the contained toxin. Boiling contaminated liquids for 15 to 20 minutes should break down the toxin. People involved in cleaning and handling the pathogen or its toxins will require use of appropriate personal protective equipment to prevent exposure.

Table 11.11 Characterization of the Ebola Virus used as a Biological Weapon of Mass Destruction.

EBOLA VIRUS	
Description	Called the Filoviridae virus, Marburg or Ebola viral hemorrhagic fevers (VHF) are caused by this virus. These viruses are believed to be disseminated by a number of potential methods. The Marburg or Ebola has the potential to be spread as an aerosol and in this regard, victims are likely to be infected by respiratory exposure.
Disease Symptoms	Initial symptoms include: • High fever, overall weakness, and severe flu-like symptoms. • Bleeding, hemorrhages from the eyes, mucous membranes, and skin. • Drop in blood pressure progressing to shock. • Failure of the blood cell-producing systems. • Kidney and liver failure • Dysfunction of the central nervous system • Death occurs for 50% to 90% of VHF victims. It is predicted that terrorist use of virus strains similar to the Marburg strain, developed as a weapon of mass destruction by the Soviet Union, would produce the highest death possible. Victims that recover from VHF infections would be vexed with kidney, liver, neurological, and heart diseases.
Preventive Treatment	None
Treatment Considerations	Decontamination of equipment and materials is recommended using hypochlorite bleach solutions and or phenolic disinfectants. Anyone providing medical support must use respiratory and total body personal protection. Aseptic procedures are required during all phases of treatment and patient support. Depending upon available resources, other more in-depth medical procedures may be available.

We have discussed considerations relating Toxic Industrial Chemicals (TICS). It has been stressed that TICS are readily available and could threaten any community. Military type chemical agents are more difficult to obtain and use, however if they are used, they would do what they were designed to do—to kill people. Microbiological pathogens, or disease causing agents, especially bacterial spore forming agents like anthrax, have been used as a terrorist threat. Viral agents are more difficult to use than bacteria when used against large numbers of people. Why? Because virus agents are obligate parasites and they require an infected living host in order to spread throughout a target population. Toxins that are produced by microbiological agents, such as botulism or ricin, can be easily used as WMD; however, because they are not contagious, they would be used as chemical agents.

In this section, we will discuss radiological, nuclear, and explosive potentials for terrorist threats.

RADIOLOGICAL DISPERSION DEVICES "DIRTY BOMBS"

What about radiological dispersion devices (RDD)? The original "dirty bomb" was developed as a nuclear bomb that contained additional materials that would increase nuclear fallout. Today the dirty bomb is a conventional explosive that contains radioactive materials that can be dispersed over a wide area without using a nuclear warhead. A RDD or dirty bomb can use waste materials for this purpose, such as wastes from nuclear reactors, laboratory wastes, and other frightening sources. It was reported in a government survey that was conducted in 2001, "since 1996, nearly 1500 pieces of equipment containing radioactive parts have been lost" ref: M.E. Marks, Emergency Responders Guide to Terrorism, p.97.

Radiological destruction to property and living things results from emissions radiation in the form of alpha, beta and gamma radiation. Alpha radiation is the least penetrative radioactive particle; beta is more penetrative to living tissues. Gamma radiation is most destructive and penetrative. Radiological harm is based upon the extent of damage and life threatening effects caused from releases of materials contaminated by these radiations. Nuclear wastes, laboratory equipment, and other devices can produce radioactive sources that may be used to create a device that can produce radiological harm.

What are sources of materials that could be used to make RDD and how can they be obtained? The following information is presented for the purpose of alerting concerned parties and recommending increased security in all applicable situations - potentials of terrorism are real:

What are Radiological Sources?

- Fission products from "cold war" nuclear devices
- Spent fuel from nuclear reactors and power plants

- Low level radioactive materials from educational and research operations, industrial and research wastes, and consumer wastes (lighted exit signs, smoke detectors, anti-static devices and many more)
- Nuclear weapons storage facilities
- Government facilities
- Medical facilities
- Industrial operations
- Dangerous goods that are being transported
- Smuggling contaminated and pure radiological dangerous goods from one nation into another nation

How Can Dangerous Radiological Products be Obtained Illegally?

- Theft from radiological resources mentioned above
- Hijacking resources of used nuclear fuel
- Procurement through means that exist within black markets located all over the world
- Espionage and bribing key personnel within any nation
- Robbery and theft from new and used dangerous radiological products while in transit anywhere in the world

What is the Value of Using Radiological Dispersal Devices for Terrorism?

- To build fear within a vulnerable people
- To inflict human injuries and death
- To destroy a nations resources
- To render lands and waters unusable

Considerations About Radiological Agents

1. Detecting the presence of a radiological agent:
   - They can be detected and measured with easy-to-use equipment.
   - Typical detection and quantification equipment includes Geiger counters, Proportional counters, Scintillation counters, and dosimeters
2. Potential use in terrorism:
   - Nuclear bomb explosion - unlikely because a device of relatively large size would be required and would be hard to obtain
   - Attack or sabotage of a nuclear reactor or other type nuclear facility - unlikely because of elevated security and international awareness
   - RDD or dirty bomb - likely because of potentials of dispersing a smaller device

3. Potential health effects from exposure to radiological agents:
   - Modes of exposure include:
     1. Penetration of human tissues by radiation
     2. Inhalation of radioactive particles into the lungs and other parts of respiratory system
     3. Ingestion or eating of contaminated foods and drinks
     4. Injection into the skin and circulatory system, as well as entry through openings in damaged skin
   - Typical acute health effects from radiological contaminants:
     1. Skin and tissue damage from burns
     2. Upper gastric and intestinal distress that would cause vomiting, diarrhea, internal bleeding, and malfunction of internal organs
     3. Loss of hair, weakness, and skin discoloration
     4. Convulsion, unconsciousness, and death
   - Typical chronic health effects from radiological contaminants
     1. Neoplasm, tumors, and cancer formation
     2. Circulatory system changes noted in blood cell formation
4. Terms Relating to Radiation Exposure:
   - (R): A commonly used term for radiation exposure is roentgen (R), which relates to exposure to gamma radiation. An exposure of 25 R of gamma radiation over a very short amount of time can result in measurable damage to human tissues. An exposure to 500 R is fatal. In some respects, roentgen is independent of the amount of time that an exposure occurs; however, if a victim is exposed to 7 R of gamma radiation during an incident and later is exposed to 10 R of gamma radiation, the exposures are accumulative and in this regard the victim has been exposed to a total of 17 R of gamma radiation.
   - (rad): A commonly used term to quantify radiation that has been absorbed is rad (radiation absorbed dose). Rates for absorbed dosages of radiation are related in rads per hour. Various substances will absorb different quantities of radiation during an exposure incident. One rad equates to absorption of 100 ergs per gram of the substance. What is an erg? A unit of energy equal to work done by a force of 1 dyne acting over a distance of 1 cm.
   - (rem): Dosage equivalent uses the term "rem" (radiation equivalent man). Dose equivalent is a quantification that indicates amounts of biological damage that occurs from radiation exposure. Most commonly used for measuring radiation dosage is millirem or one thousandth of a rem. Dosage in rads is converted to rems by using a factor of 1 for beta and gamma radiation, and a factor of 10 for alpha radiation. In general, standard national guidelines recommend dose limits of 0.5 rem per year. International guidelines recommend limits of 0.5 rems per year for short-term exposure and 0.1 rems per year for long-term exposure. Exposure rates, or radiation exposure in units of time, are usually presented as

roentgen per hour (sometimes as milli-roentgen per hour). The rate of radiological exposure is usually presented to denote hazard levels at the source of radiation.

- (curie): A unit of radiation activity that indicates the disintegration rate of a radioactive product, also called decay, is called curie (Ci). Hazardous characterizations for radioactive materials are quantified as low activity or high activity - the later being most dangerous. Levels of dangerous radioactive properties are usually presented in terms of curies per gram (Ci/g) of radioactive material.

5. Examples of radiation exposure effects:
   - 1000 rem - Illness and eventual death within 3 weeks
   - 100 rem - Occurrence of radiation sickness
   - 5000 millirem (or 5 rem) - The lowest known dosage that indicates evidence of cancer in adults; also is the highest one-year occupational dosage allowed in the United States
   - 300 millirem (or .3 rem) - Typical annual background radiation dosage from natural sources in North America
   - 30 to 60 milirem (.03 to .06 rem) - Example of annual dosage from artificial radiation sources such as seen in medical testing

6. Radiation factors that have an affect upon human health:
   - Total amount of radiation exposure.
   - The amount of time that living tissues are exposed to radiation.
   - The source and kind of radiation.
   - Portions of the human body that are exposed.
   - Biological variables such as: condition of health, age, gender etc.

7. How to be protected from radiological hazards:
   - ALARA: Prevent radiation exposure by following safe practices and ensuring that exposure potentials are As Low As Reasonably Achievable.
   - Follow recommended safe practices such as those recommended by the Nuclear Regulatory Commission and other authorities, such as: TDS
     1. Keeping potential exposure times to very short durations: T = TIME
     2. Staying at safe distances away from potential radiation hazards: D = DISTANCE
     3. Use protective shielding from potential radiation sources: S = SHIELDING
     4. Use appropriate personal protective equipment
     5. Following safe work practices, standard operating procedures, and other proven principles

In summary, three principles are emphasized when efforts are being made to control electromagnetic radiations that can penetrate the human body (also called ionizing radiation):

1. Keeping lengths of exposure time to minimum amounts
2. Staying as far away from the radiation source as possible
3. Use shielding as protection from radiation sources

The amount of radiation exposure that a person may be exposed to is relative to the rate of dosage and the amount of time that the person may have been exposed. It must also be noted that when the distance from a radiation source is doubled, the exposure is reduced by a factor of 4. In this regard, if a person moves twice the distance from an original exposure distance, the potential for exposure drops to ¼ of the original amount of exposure.

Also, alpha and beta radiation exposure can be radically reduced and even stopped when people are shielded in vehicles, buildings, and within relatively dense structures or materials. Gamma and neutron radiation requires extremely dense shielding, such as thick concrete, lead, and other dense materials.

NUCLEAR BOMBS AND EXPLOSIVES

At this writing, the following nations have been declared to be nuclear powers:

- United States of America having 10,640 nuclear weapons
- Russia having 8,600 nuclear weapons
- People's Republic of China having 400 nuclear weapons
- France having 350 nuclear weapons
- United Kingdom having 200 nuclear weapons
- India having 90 nuclear weapons
- Pakistan having 48 nuclear weapons

In 1985 there were believed to be over 65,000 nuclear weapons worldwide. It was reported in 2002 that at least 40,000 nuclear weapons still remain throughout the world. (Reference: http://www.nrdc.org/nuclear/nudb/datainx.asp)

Other nations that are known, and some are suspected, to have had nuclear weapon programs and at this time claim to have reduced or stopped developing nuclear capabilities. Some of these nations include: Israel, Iran, North Korea, and Ukraine.

Other nations that were known to have developed nuclear capabilities but are now regarded as no longer active in nuclear weapons development include: Argentina, Australia, Belarus, Brazil, Egypt, Germany, Iraq, Japan, Kazakhstan, Libya, Romania, South Africa, South Korea, Sweden, Switzerland, and Taiwan. Other nations capable of developing nuclear weapons include: Canada, Netherlands, and Saudi Arabia.

COMMON NUCLEAR WEAPONS THAT HAVE BEEN DEVELOPED

Two major configurations of nuclear bombs have been developed over the past 60 + years:

1. Nuclear fission bombs or atomic bombs such as those dropped upon Japan to end World War II
2. Nuclear fusion bombs or the hydrogen bomb

Characteristics of Nuclear Bombs

- Atomic or A-Bombs are nuclear fission bombs that are the result of contained radioactive metals such as uranium or plutonium that are exposed to streams of flowing neutrons. The emitting neutrons cause the radioactive metals to split into lighter elements that emit more neutrons. The increased emissions of more and more neutrons cause chain reactions that split (or cause fission) within the nuclei of all materials near the degrading radioactive materials. As atomic fission occurs, extreme amounts of all energies are released from the surrounding environment.
- Hydrogen or H-Bombs are nuclear fusion bombs that occur as a result of light weight atoms such as hydrogen and helium that are directed by physical forces to combine into heavier atoms. When this occurs, extreme amounts of energy are released into the immediate environment. These bombs are also called thermonuclear weapons because in order to trigger nuclear fusion chain reactions, extremely high temperatures must be produced. In order to produce enough energy for a fusion reaction to occur, a fission bomb is used to start the chain reaction.

Over the years, nuclear weapons have been characterized as fission or fusion devices primarily because of the nature of the energy sources used to make the bomb. However as mentioned above, fission and fusion is used to produce the most powerful bombs. In this regard, most scientists no longer use the terms fission or fusion device. The term now used is nuclear weapons of mass destruction.

DIRTY BOMB—CONSIDERED TO BE A NUCLEAR WEAPON? NO.

As mentioned earlier in this chapter, "Dirty Bombs" are considered to be RDD Weapons, or radiological dispersion devices. RDD's are non-nuclear explosives that contain radioactive materials that are packed with a conventional explosive and, as it explodes, it scatters radioactive contamination throughout a populated area. Relatively easy-to-assemble weapons are an everlasting nightmare that people in all nations fear throughout today's confused world. Along with causing serious human health hazards and death, a dirty bomb can render lands and resources un-inhabitable for many years after explosion.

MODERN THERMONUCLEAR WEAPONS

Nuclear warfare has evolved to weapons that have increased destructive forces many times greater than bombs we have already mentioned. One property that is used involves placing an enriched fissionable uranium shell around fusionable materials. Thus, the worst scenarios of both destructive systems are combined.

Isotope Salted Bombs: Cobalt has been introduced into nuclear fusion bombs. During the fusion reaction, cobalt metal is converted to cobalt 60 which becomes a powerful emission source of gamma rays that produce long term radioactivity dangers for at least 5 years. This type of bomb is called an isotope salted bomb and various substances are added to adjust hazard times that remain (fallout) after the bomb has been detonated. Salting has been accomplished using gold for short term fallout periods (few days). Tantalum and zinc isotopes have been used to produce hazardous fallout dangers that remain for several months. These types of bombs create extremely radioactive fallout that contaminates land that people cannot inhabit for long periods of time.

THE NEUTRON BOMB, AN ENHANCED RADIATION WEAPON

These are relatively small thermonuclear weapons that are designed to release emissions of neutrons into the environment at detonation during the fusion reaction (normally the neutrons are contained within the fusion reaction). At detonation an intense burst of high energy neutrons will penetrate almost everything in their path (including most common shielding materials) which can cause death to people that have found shelters and shielding.

WHAT HAPPENS DURING A NUCLEAR EXPLOSION?

This following topics are presented to characterize events that usually occur when a nuclear device is detonated:

- Energy release
- Blast effects
- Electromagnetic Thermal Radiation (ETR)
- Electromagnetic Pulse
- Radiation
- Fallout

Information listed in Table 12.12 describes how energy is released when a nuclear weapon is detonated.

FOUR CHARACTERIZATIONS OF ENERGY RELEASED DURING A NUCLEAR EVENT

Weapon design and area detonate contribute to the distribution of energy released. Fallout or residual radiation is considered to be latent or delayed energy release.

Conventional Explosives vs. Nuclear Explosives

The prime difference between conventional explosives and nuclear explosives is the magnitude of damage and residual dangers that nuclear explosives produce. When a nuclear device is detonated, all initial reactions occur within a millisecond. During the initial blast, energy is emitted primarily as thermal and light (x-rays) radiation. Most of the energy heats the air and forms a fireball. The majority of remaining energy is transferred to the kinetic energy of rapidly moving debris. As the environment surrounding the blast may be dense, e.g. solid materials on the ground, water, buildings etc., the shockwave will be strong. If the nuclear device is detonated at high altitudes, where air density is low, electromagnetic energies will travel long distances before being absorbed.

Blast Effects or Air Bursts

When a nuclear bomb is detonated, winds from the blast will exceed several hundred miles per hour. The blast range is proportional to the explosive powers within the weapon. There are two specific phenomena that are related to the blast wave and the air:

1. Air density vs. shock wave pressure: Overpressure is directly proportional to the air in the shockwave. In this regard, there is a sharp increase in pressures that are contained in the shock wave.

Table 11.12 How energy is released when a nuclear weapon is detonated.

Source of Energy Release	Distribution
Detonation	40 to 60 % of Energy Released
Thermal Radiation	30 to 50% of Energy Released
Ionizing Radiation	5% of Energy Released
Residual Radiation or Fallout	5 to 10% of Energy Released

2. Winds form that travel away from the blast and the pressures of the winds cause destruction to anything in its path.

In combination, high static compressed overpressures of the blast wave and the blast winds cause most material damage during a nuclear explosion. Physical forces related to air compression, vacuum, and drag, form blast effects that radically exceed forces seen in the strongest hurricanes.

Electromagnetic Thermal Radiation (ETR)

Electromagnetic radiation (EMR) is emitted during a nuclear detonation in the forms of visible, ultraviolet, and infrared light. The impact of the ETR is dependent upon the weapon size. The primary damage to humans is burns and eye damage. The intensity of ETR during a nuclear explosion will cause fires in debris that are spread by the blast. Thermal radiation travels in straight lines as it is emitted from the nuclear fire ball. If fog or haze causes the emitted light to scatter, everything will become superheated and, in this regard, protective shielding is not effective. Bright reflective substances can beam the destructive light away and dull or opaque materials will be destroyed by extremely high temperatures. During the bombing of Hiroshima, Japan, a large firestorm occurred within 20 minutes of detonating the atomic bomb. A firestorm consists of strong winds that blow inward toward the fire source.

Electromagnetic Pulse (EMP)

During a nuclear explosion, gamma rays produce high energy electrons that are captured by earth's magnetic field. This occurs at high altitudes and causes an electronic disturbance called an electromagnetic pulse. The initial EMP may last about 1/1000th of a second, and secondary EMP may follow lasting slightly longer. Electromagnetic pulse can destroy unshielded electronic devices, radio transmitters, and communication systems. A device that is detonated at high altitudes could cause widespread electronic malfunctions including stoppage of ground transportation vehicles, some aircraft, and communication systems.

Radiation

During an initial nuclear air burst, approximately 5% of energies that are released are in the form of neutron and gamma radiation. Both fission and fusion reactions emit neutron radiation and initial gamma radiation will occur from this source, as well as from radioactive decay that occurs in short-lived fission reactions. Initial radiation decreases as it travels from its source. Radiation reduction also occurs by environmental and atmospheric absorption and by scattering. Near the point of detonation neutron density is higher than gamma ray intensity. As distance increases from the point of detonation, the neutron density diminishes.

Nuclear Fallout

Residual contamination by nuclear products and neutron-induced activity occurs well after a nuclear detonation. Depending upon weather and wind movement, larger contaminated particles may start falling from the skies within 24 hours. Very light particles can rise as high as the stratosphere and be sprinkled globally during weeks to months that follow. The following are examples of materials that can be included in nuclear fallout:

- Fission Materials - There are over 300 different types of contaminated fission materials that may be found after nuclear detonations. Some may be hazardous for very short periods of time and some may last months and even years. They usually decay to produce emissions of beta and gamma radiation, both penetrate human tissues and cause cellular damage.
- Fissionable Materials - materials that have not completed the fission reaction may be found in nuclear fallout. This may include uranium, plutonium, and other materials used for enrichment or salting. Products of these materials may likely form alpha radiation hazards that are relatively low hazards, unless ingested or subjected to human mucosal tissues where serious illness may occur.
- Neutron-Induced Hazards - materials may become radioactive by neutron induced actions and then decay over extended periods of time yielding beta and gamma emissions that penetrate human tissues and cause severe damage. Neutrons that are emitted during the initial detonation reaction will activate weapon residue materials. Environmental materials such as soil, air, and water can also be activated, depending upon their location from the initial detonation.

CONCLUSION

Dangers of nuclear attack are real and unfortunately will remain, as long as greed infests the nature of mankind. People that walk the earth today have witnessed and, during the new millennium, will most likely witness the extended horrors of terrorism. Weapons of mass destruction have been produced and very likely as you read this book are being produced.

Dear readers regardless of where we may live we must find a way to do what we can to stop terrorism, and to encourage the entire world to stop building weapons that will destroy the dreams of our brothers and sisters across this planet.

Whatever we do or choose not to do will be acknowledged by our posterity.

ACKNOWLEDGEMENT

The author would like to thank all the personnel at Kirkwood Community College, HMTRI, Cedar Rapids, Iowa who is part of the Midwest OSHA Education Centers, who through their hard work and professionalism, developed and presented the OSHA 5600 Site Worker Train-the-Trainer & NIEHS Weapons of Mass Destruction Disaster Site Worker Safety Train-the-Course. Much of the information contained in this chapter reflects the outstanding training and valuable information received by the author while attending course work at Kirkwood.

Also, the author would like to acknowledge the following resources that also had an influence upon the works contained in this chapter:

Mr. Michael E. Marks	*Emergency Responder's Guide to Terrorism* by Red Hat Publishing.
FEMA	*Hospital Emergency Response Training (HERT) for Weapons of Mass Destruction (WMD) Events* by the Federal Emergency Management Agency.
George Buck, PhD	*Preparing for Terrorism, an Emergency Services Guide* By Delmar, Thompson Learning

GLOSSARY

PURPOSE

The following are basic terms and definitions that are common to the environmental, health, and safety profession. This glossary is not intended as a legal or scientific reference.

Abatement - Actions taken to reduce the degree or severity of chemical or biological contaminations.

Absorbent - Materials that are designed to collect or contain substances using principles such as adsorption, absorption, stasis, and others.

Absorption - Exposure to hazardous substances by transfer directly through the skin.

Acceptable Risk - A consideration, after evaluating negative potentials and weighing corresponding positive benefits, determined to be a hazardous task that is worth performing.

ACGIH - American Conference of Government Industrial Hygienists. An organization that focuses upon occupational health and safety. ACGIH develops and publishes recommended hazardous substance exposure limits.

Acids - Substances with low pH values (0-6) and are corrosive. They have potentials to destroy living tissues, dissolve metals and other substances. RCRA regulates hazardous waste acids at pH 2 or less.

ACS - American Chemical Society

Acute Toxicity Effect - Injury or illness manifestation within a short time after exposure to a hazardous substance.

Adsorption - Molecular adhesion (sticking) that facilitates filtering organic substances found in liquids or vapors, when passing them through carbon filters or activated charcoal.

AEC - Atomic Energy Commission - In 1975 regulatory responsibility was given to the Nuclear Regulatory Commission. Weapons production and research activities were given to the Energy Research and Development Administration.

Aerosol - Fine particles of substances that diffuse into air and settles as sedimentary layers within the environment.

Air-Line Respirator - Continuous air supplied from a compressed air source.

Air Monitoring - Using test equipment to determine atmospheric pollution.

Air Purifying Respirator (APR) - A respirator that uses chemicals or filters to remove particulate matter and vapors. APR's are limited for use in atmospheres containing greater than 19.5 % oxygen and contaminant concentrations are within rated levels for the specific device.

Alkalies - Substances with pH values >7 (caustics, bases, hydroxides).

Allergen - A substance capable of causing an allergic reaction.

Allergy/ Sensitivity - A reaction to a substance that has an affect upon the immune system such as overformation of antibodies. (Dermatitis, anaphylactic shock, hives).

Alloy - Mixtures created by melting metals together or by combining polymers and polymer compounds.

Alpha Particle - Large radioactive particle has a mass value of 4 amu and is the size of a helium atom. This particle can be encapsulated by a sheet of paper or plastic. Can be destructive to internal human and animal tissues.

AMU - Atomic Mass Unit.

Anhydrous - Free of water content, dry.

Anion - Ions with a negative (-) charge.

ANSI - American National Standards Institute, a private organization that develops standards for equipment and recommends safe practices for hazardous operations and others.

Antidote - A substance that will stop or reverse the effect of a toxic substance.

Appearance (Chemical) - A physical description of a substance that includes color, liquid, solid, vapor, gas, morphology, odor, and consistency.

APR - Air purified respirator, a chemical cartridge respirator. Use requires proper fit, training, and the user must be medically fit to wear it. Your instructor will discuss limitations and liabilities for using this kind of respirator.

AQTX - Aquatic Toxicity

Assessment (EHS) - Protocols for evaluating environmental and safety hazard potentials, mitigating conditions and situations that may effect regulatory compliance or better yet "common sense."

Asphyxiation - Tissues of a living system that cannot receive oxygen, due to chemical reaction or from physical blockage of the respiratory system.

ASTM - American Society for Testing and Materials.

Asthma - Constriction of the airways to the lungs, producing symptoms of cough and shortness of breath, can be related to allergic reaction.

Asphyxiant - A substance that can cause unconsciousness or death due to the lack of oxygen.

Atmospheric Testing - Above ground testing of nuclear devices.

Atom - Smallest part of matter that can be identified as a specific element. Atoms consist of protons (+) and neutrons (0) in a nucleus which is surrounded by electrons (-).

Atomic Number - The number of protons in the nucleus of an atom.

Atomic Weight - Relative weight of an atom which consists of protons and neutrons in the nucleus.

Auto-Ignition Temperature - A temperature that a substance will burn without external ignition *sources*.

Back-Draft - An explosion caused by the influx of air into heated mixtures of gases. At least one gas in the mixture is heated above the ignition temperature.

Base - A substance that contains hydroxide ion (OH), see alkali.

BEI - Biological exposure index, the maximum recommended value of substance in body fluids, tissues, or exhaled air.

Beta Particle - Emission of radiation by decay, a reaction seen in many radio nuclides. A beta particle is identical to an electron and has a short range in air with low penetration potential.

Binary Compound - Chemical that contains only two elements.

Bio-Accumulation - Buildup of contaminants within a living system.

Bio-Aerosols - Airborne microbial by-products, spores, and other microbiological agents capable of causing disease or materials destruction.

Biodegradable - A substance that has the potential to be consumed or chemically altered by living things.

Bio-Hazard - Biological sources of hazards such as pathogenic organisms and hazardous by-products generated by micro-organisms.

Biological Half Life - The time taken for a living system to eliminate one half the dosage of a hazardous substance.

Blasting Agent - Materials that are not usually considered explosives that are relatively insensitive to heat and shock. These materials usually require special procedures to cause them to explode (Ammonium Nitrate + Fuel Oil).

BLEVE - Boiling Liquid Expanding Vapor Explosion.

Blood Asphyxiant - Substances that interfere with the ability of red blood cells to carry and release oxygen to the tissues in the body.

Blood Borne Pathogens - Microbial disease causing agents found in blood and body fluids (HIV, Hepatitis, and many other pathogenic organisms).

Body Burden - Total quantity of radio nuclide in the body.

Boiling Point - The temperature at which a liquid becomes a vapor at a given pressure.

Bonding - Chemical bonding involves attachment of elements (atoms) to form molecules by ionic, electrovalent, attraction, and by covalent, sharing of electrons.

Breakthrough Time - The amount of time required for a given substance to penetrate or degrade personal protective equipment.

Brisance - "Explosive Power," the surrounding effects of an explosion.

Buddy System - A requirement for workers in hazardous materials environments to never work alone, have back up support and to observe their coworkers.

Buffer - Substances that reduce changes in hydrogen ion concentration, a mechanism of pH control.

Carboy - A chemical container often considered a special container that may be reused and may be reinforced for sound handling.

Carcinogen - A substance capable of causing cancer.

Cartridge Change Schedule - Air purified respirator cartridge replacement schedule that is established by the employer and is based upon manufacturer and OSHA recommendations.

CAS Number - An internationally recognized numerical, chemical identification system often seen in MSDS and other references (Chemical Abstracts Service).

Cascade System - Equipment used to refill air tanks with Grade D Air as used with self-contained breathing apparatus respirators.

Catalyst - A substance that causes a change in a chemical reaction, but it is not changed by the reaction.

Caustic - An alkali that causes strong irritation, corrosion of metals, and can destroy living tissues.

Ceiling - Maximum exposure limit to dangerous substances "not to be exceeded."

Cell - The structure unit that living tissues are made of. There are many types of cells such as nerve cells, muscle cells, blood cells, and skin and mucosa cells (each type of cell performs a specific function).

Central Nervous System - Includes the brain and spinal cord, which controls activities for the entire nervous system.

Centigrade (also call Celsius) - International temperature measurement such that water boils at 100 degrees centigrade (1 Atm) and freezes at 0 degrees centigrade.

CERCLA - Comprehensive Environmental Response Compensation and Liability Act. The Federal law that created funding for cleaning up abandoned and uncontrolled hazardous wastes sites, SARA/Superfund.

CFR - Code of Federal Regulations: 40 CFR = Environmental; 29 CFR = Safety and health; 49 CFR = Transportation

Chain of Custody - Protocols for proper handling of chemical or biological test samples for the purpose of ensuring accuracy, integrity, and to eliminate potentials for special interest alteration of test results.

Chemtrec - Chemical Transportation Emergency Center, operated by the Chemical Manufacturers Association, a service for emergency support 1-800-424-9300.

Chemical Compound - Proportional arrangement, by weight, of two or more elements such that the properties differ from the individual elements.

Chemical Family - Elements or compounds that have a common general name.

Chemical Protective Clothing (CPC) - clothing manufactured for the purpose of protecting workers that handle dangerous substances.

Chemical Reaction - Change of chemical composition and properties by forming new bonds and structures.

Chemical Separation (nuclear reprocessing) - A method for extracting uranium and plutonium from spent nuclear products. Extracted fission products are high level wastes.

Chemical Name - Scientific designation of a substance that may include listing by the International Union of Pure and Applied Chemistry (IUPAC) or the Chemical Abstract Service (CAS).

Chemistry - The science of matter, energy, and reactions.

Chromosome - A portion of a cell that contains genetic materials.

Chronic Effect - Exposure to hazardous substances causing adverse effects upon living systems that may not manifest symptoms for long periods of time.

Clandestine Lab - A lab operated for the purpose of making illegal substances.

Clean Air Act (EPA) - Federal law enacted to regulate, clean up, and prevent air pollution.

Clean Water Act (EPA) - Federal law enacted to regulate, clean up, and prevent water pollution.

Cleveland Opens Cup - A method for determining the flash point of an ignitable liquid.

Colorimetric Tubes - Glass tube test devices that are used to determine quantities of air borne contaminants. Each tube is specific to a substance and contains a chemical media that turns color to indicate the presence of a contaminant as air is pumped through the tube.

Combustible - A substance that has a flashpoint temperature that exceeds 100 degrees F.

Common Chemical Name - Chemical identification such as a code name, number, trade or brand name, and generic name(s) other than the chemical name.

Common Sense - Sound judgment based upon knowledge and obvious variables.

Community Right to Know - Laws that require business operators and property owners to provide information for the emergency community and for public record, that declares storage and usage of hazardous or otherwise regulated materials at a given location.

Compressed Gas - A gas or mixture of gases in a container with pressure of at least 40 pounds per square inch at 70 degrees Fahrenheit; or a gas or mixture of gases in a container with an absolute pressure exceeding 104 psi at 130 degrees Fahrenheit; or a liquid with a vapor pressure that exceeds 40 psi at 100 degrees Fahrenheit.

Computer-Aided Management of Emergency Operations (CAMEO) - A database provided for the purpose of supporting emergency planning and hazardous materials emergency response operations.

Concentration - The amount of a given substance as compared to its mixture ratio with other substances.

Confined Space - A space with limited access that is not normally inhabited and is difficult to enter and exit. A permit-required space is a confined space that contains hazards or has associated hazards that may affect employee safety while working within.

Consignee - The location or address that materials are being shipped to.

Container (HAZMAT) - A device that holds hazardous materials during shipment, storage, or during use.

Containment - Encapsulation, encasement, or method of stabilizing operations that may potentially emit or release hazardous materials, chemicals, or biohazards into the environment.

Contamination - Deposition of hazardous substances on surfaces, people, structures, and objects.

Contamination Reduction Zone (CRZ) - the location of the decontamination corridor in the zone following the exclusionary "hot" zone and entering the support "cold" zone that surrounds a controlled hazardous materials site.

Corrosive - A substance having potential to destroy human and plant tissue, or to dissolve metal.

Covalent Bonding - Chemical bonds involving shared electrons.

Cracking - Breaking apart chemical bonds.

Cryogenic - Hazardous substances with extremely cold temperatures, as boiling cryogenic gas (< -150 degrees F) that can destroy human, animal or plant tissues.

CPSC - Consumer Products Safety Commission.

Cubic Meter - A standard measure of volume, usually related to concentrations of substances in air. One cubic meter is equal to 35.3 cubic feet, 1.3 cubic yards, or one million cubic centimeters.

Cutaneous Hazards - Chemicals that damage or irritate the skin.

Curie - A measure of radioactivity. 1 curie = 37 billion radioactive decays per second.

Cyanosis - Blue coloring of the skin, indicating the lack of oxygen.

Damage Assessment - Obtaining and evaluating information relative to the extent and cost of damage related to an incident.

Daughter - A nuclide formed by radioactive decay or by a decay product.

Decay (Radioactive) - Spontaneous emission of particles and energy from the nucleus of an unstable atom.

Decomposition - Chemical breakdown of materials into simple parts such as compounds, elements, or atoms.

Decon (decontamination) - Removal and disposal of hazardous pollutants.

Deep Well Injection (Geologic Repository) - disposal of waste products deep beneath the earth's surface.

Deleterious Substances - Materials not obviously harmful to humans, but may have a harmful effect upon the environment.

Density - Mass or weight per unit volume of substance, D= M/V.

Dermal - Relating to the skin.

Dessicant - A substance that can extract and collect moisture from the air.

Dike (Embankment or ridge) - Put in place for the purpose of containment and controlling the flow of liquids and other material forms.

Dispersion - To spread materials by scattering, diffusing through air, mixing into soil, or by being carried by water or other mobile mechanisms.

Diversion - Controlled movement of hazardous materials to locations that may least likely be affected by dangerous material properties.

DOE - Department of Energy was created from the Research and Development Administration and other Federal functions in 1977. It is responsible for production of nuclear weapons, energy research, and clean up of waste sites that are under its authority.

Dose - A specified amount of radiation or hazardous substance that is absorbed by a living system. (Usually expressed written in milligrams of substance per kilogram of body weight of test animal or victim, mg/kg).

DOT - U.S. Department of Transportation

DOT Proper Hazardous Shipping Name - Hazardous material name listed in 49 CFR 172, table 101. Includes hazard class, United Nations Number or North American Number, Packing Group Number, RQ, and other significant information.

Dust - Finely divided solid properties of materials that have a potential to become airborne and eventually settle to contaminate other materials.

EHS Audit & Assessment - An official evaluation of a site that uses specific protocols for determining environmental pollution potentials, worker safety potentials, and status of regulatory compliance.

Edema - Swelling of living tissues due to water or fluid accumulation.

Epidemiology - The study of a pattern of disease that may spread within a population of humans or animals.

Element - A substance composed of entirely one kind of atom. Elements are designated by chemical symbols. (Cu = copper, Au = gold, Ag = silver, Hg = mercury, Pb = lead).

Emergency Response - Actions taken to stabilize dangerous conditions.

End-of-Service-Life Indicator - A system associated with cartridges used on air purified respirators that indicates when a cartridge is ready to be removed and replaced.

Endothermic Reaction - A chemical reaction that absorbs heat.

Engineering Controls - The first choice to consider before donning respiratory protection. Engineering controls may include work area modification, ventilation, and other alternatives to wearing a respirator.

Environmental Management - A Department of Energy organization that is responsible to oversee nuclear environmental clean up efforts.

Environmental Contamination - Release of hazardous substance that causes pollution to the environment.

EPA (Environmental Protection Agency) - 1970, Federal agency responsible for enforcing environmental laws.

Epidemiology - The study of disease sources and distribution of disease and injuries in human populations.

Etiological Agent - A disease-causing or toxic substance associated with contamination generated by a micro-organism.

Evacuation - An orderly removal of personnel from potentially dangerous conditions to safe locations.

Evaporation - A process of changing liquid to a vapor and mixing it into the surrounding environment.

Evaporation Pond - An aqueous liquid waste holding area that becomes concentrated as water is evaporated.

Evaporation Rate - Based upon ethyl ether = 1; Fast Evaporators are > 3, Medium Evaporators are .8 to 3.0, and Slow Evaporating Substances are <.8.

Excavation - A manmade cut or depression into soil that requires following specific regulatory guidelines and protocols to guarantee safety of individuals that may enter.

Exclusionary Zone - The contaminated or "hot" zone within a controlled hazardous materials site.

Exotermic Reaction - A chemical reaction that liberates or releases heat.

Explosive Limits - Concentration ranges of substances mixed in air that can result in an explosion or fire.

Fahrenheit - Temperature in degrees (F) such that 212 degrees F at atmospheric pressure (760 mm Hg) will boil water and 32 degrees F will freeze water.

First Responder (See 29 CFR 1910.120 q.) - The first authorized individuals on scene at a hazardous materials incident. The OSHA regulation defines responsibilities of first responders at the awareness, operations, technicians, specialist levels, and other associated competent persons.

Fission - Splitting apart the nucleus of a heavy atom. Usually caused by absorption of a neutron, results in the liberation of large amounts of energy and at least one neutron.

Flammable - Substances that will readily ignite and burn vigorously.
Liquid = Flash point <100 degrees F (OSHA)
Solid = Defined in 29 CFR 1910.109 (e), potential to cause fire and burn vigorously.
Gas = Forms flammable mixtures with air, at less than 13% concentration. Note: DOT requires ignitable liquids with flash points of 140 degrees F and lower, to be labeled as flammable.

Flammable (explosive) Limits - Lower explosion limit (LEL): Minimum concentration of a substance allowing it to ignite or explode. Upper explosion limit: concentration of a substance that will "flood out" and prevent burning or explosion.

Flash Point - The lowest temperature at which a substance will emit enough vapors to burn or explode when an ignition source is present.

Fume - Airborne dispersion of hazardous emissions usually related to heating of metals and other solids to temperatures of sublimation (short term vapor state of matter).

Fume Fever - An acute condition caused from exposure to metal and metal oxide fumes, an illness common to welders that do not ventilate or use respiratory protection.

Fusion - A reaction that occurs when the nuclei of lighter elements combine to form the nucleus of a heavier element, which releases large amounts of energy.

Gamma Radiation - High energy emission of deeply penetrating electromagnetic radiation from the decay of radio nuclides.

Gene - A part of a chromosome that carries a particular inherited characteristic.

General Exhaust - Ventilation of air contaminants from a general work area.

Gram - A metric measure of weight, 28.4 grams = 1 oz.

Grounding - Attachment of wire to equalize electrical potential between at least 2 electrically conductive containers.

Half Life - The time required for one half of a given number of atoms to disintegrate (decay). Every isotope has its own half life properties; some may last only a fraction of a second, and others lasting billions of years.

Halogen Chemicals - Containing Chlorine, Fluorine, Bromine, and Iodine (halons).

Hazard Class - Nine (9) classifications of hazardous materials as designated in the DOT 49CFR regulations of dangerous goods shipping, transportation, and storage.

Hazardous Material - Substances that have properties capable of causing harm or damage to living systems, or to the environment.

Hazardous Ingredients - Constituents of a mixture that by themselves may cause injury or damage to living systems or to the environment.

Hazardous Substances - Synonymous with hazardous materials for most practical purposes. The Department of Transportation considers a hazardous material to be a hazardous substance when the amount of material, in one container or one shipment, equals or exceeds the Reportable Quantity (RQ).

Heat Stress - An unsafe condition that occurs when workers are subjected to elevated temperatures, especially during the use of full coverage personal protective equipment. Symptoms include profuse sweating, cramps, and exhaustion. Uncontrolled heat stress is a precursor to heat stroke, a condition that can be fatal.

Heavy Water - Water that has deuterium atoms in place of hydrogen atoms.

Hematopoietic System - The mechanism of forming blood in the human body.

Hepatotoxin - Substances that damage the liver.

High Level Waste - A hazardous waste that contains highly radioactive products or other concentrated hazardous properties.

Hydrogen - The simplest element or atom. (Two isotopes of hydrogen, deuterium and tritium are used in nuclear weapons).

Hydrocarbons - The building blocks of organic substances composed primarily of hydrogen and carbon.

Hygroscopic - Moisture-absorbing substances.

Hypergolic - Spontaneous ignition when two or more chemicals are mixed.

IDLH - Immediately Dangerous to Life and Health, an environment that is very dangerous due to high concentrations of toxic substances, insufficient amounts of oxygen, and or other conditions.

Ignition Temperature - The lowest temperature that a substance will sustain combustion and burn.

Immiscible - Substances that will not mix together.

Incident (Hazmat) - An event that involves release or potential release of dangerous substances into the environment.

Incident Command - Coordination and management of all activities directed at stabilizing emergency conditions during an incident. The incident commander or senior response official is responsible to manage emergency response actions.

Incompatible - Substances that produce dangerous reactions when mixed.

Inert Gas - A gas that does not react with other substances.

Ingestion - Eating or taking internally, a mode of exposure to hazardous substances.

Inhalation - Breathing in—one of the fastest modes of exposure to hazardous substances.

Inhibitor - A chemical additive used to prevent unwanted changes in a chemical reaction.

Initiator - A substance that starts a chemical reaction.

Injection - A mode of exposure to hazardous substances by entering the body through broken skin.

Ionizing Radiation - Electromagnetic radiation that has potentials to change atomic charge and bond properties usually includes alpha, beta, gamma, and other radiations.

Irradiate - Exposure of materials to radiation.

Irritant - A substance that causes inflammation to living tissues.

Isotopes - Forms of an element that vary by the number of neutrons in the nucleus.

Kilogram - 1,000 grams or 2.2 pounds.

Lab Pack - Generation of regulated wastes by accumulating them in small quantities and shipping them to a licensed treatment, storage, and disposal facility.

Latent Period - The period of time between exposure and the onset of symptoms.

LC50 - (Inhalation Hazards) Lethal concentration of a substance that causes 50% of test animals to die.

LD50 - (Injection or Ingestion Hazards) Lethal dosage of a substance that causes 50% of test animals to die.

LEL - Lower explosion limit, minimum concentration of an ignitable or explosive substance that will burn when exposed to an ignition source.

LEPC - Local emergency planning committee, an organization within the community that is mandated by SARA Title III regulations to plan for and monitor emergency actions.

Liter - Metric unit for volume. (One U.S. quart is about .9 liter; One liter contains 1,000 cubic centimeters).

Local Exhaust - Removal of air contaminants through ventilation located near the point of generation.

Manifest - Documents that accompany hazardous waste shipments and are used by waste generators and regulators to monitor waste management and waste generation activities.

Marking Containers - Requirement of OSHA Hazard Communication regulations to appropriately mark all containers of hazardous materials. Labels should contain information such as: name of substance, specific hazards, and target organs affected, methods to prevent exposure, manufacturer name, first aid instruction, suggested PPE, and other information.

Melting Point - Temperature that a solid substance becomes liquid.

Metabolism - Chemical activities within a living system to convert substances into energy.

Micro-Organism - Virus, bacteria, fungi, and other microbiology that may cause disease or may present health and toxicity conditions in humans, animals, and plants.

Milligram (mg) & **GRAM** (g) - Metric unit of mass or weight (One gram = 1,000 mg, one U.S. ounce = 28.375 grams or 28,375 mg).

Mg/Cubic Meter - Concentration of substance mixed in air (mass/volume of air).

Mg/L - Concentration of substance equal to 1 part per million. (1/1,000 gram per liter gas or liquid measure).

Millimeters of Mercury (mmHg) a common measure of pressure. (At sea level, earth's atmospheric pressure is 760 mmHg).

Mitigation - Actions used to contain, reduce severity, or to eliminate harmful potentials related to hazardous materials spills or releases into the environment.

Mixed Waste - Waste that contains radioactive and other hazardous properties.

Modes of Exposure - Four ways people become exposed to dangerous substances: inhalation, ingestion, absorption, injection, or entry into cuts and open wounds.

Molecule - Two or more atoms that are bound together.

Monomer - Small organic molecules that combine to form larger molecules (polymers). Rapid polymerization is a hazardous reaction that involves the release of large amounts energy.

Monitoring - Checking potentials for pollution or contaminant levels in the environment.

MSDS - Material Safety Data Sheet, chemical hazard and other information provided by the manufacturer of a chemical product.

MSHA - U.S. Mine Safety and Health Administration

Mucosa - Human and animal tissue that is a common site for exposures to hazardous substances (soft tissue, eyes, nasal, genitourinary, and other tissues).

Mutagen - A hazardous substance or electromagnetic radiation that has the potential to alter the genetic properties of living cells.

Narcosis - Effects upon the human body from exposure to hazardous materials that may cause fatigue of the nervous system, stupor, or unconsciousness.

Natural uranium - Non-enriched uranium, it consists of 99.3% uranium 238 and .7% uranium 235.

Necrosis - Death of living tissue caused by exposure to hazardous substances.

NEPA - National Environmental Policy Act, 1970, a law that requires the federal government to consider environmental impacts in the decision making process.

Nephrotoxins - Substances that can cause kidney disease and organ damage.

Neurotoxins - Substances that have a negative effect upon or cause damage to the nervous system.

Neutralize - To stop the corrosive properties of an alkali or acid by reacting an acid with a weak base, or by reacting an alkali with a weak acid. Neutral pH is considered to be between pH 6 and 8.

Neutron - A large uncharged particle in the nucleus of an atom.

NFPA - National Fire Protection Association, an association that helps develop fire fighter safety procedures. It is widely known for developing the 704 hazardous substance emergency placard system.

NIOSH - National Institute for Occupational Safety and Health, a government agency that assists OSHA in setting safety and health standards through research and testing.

NRC - National Response Center for hazardous substance spill and release notification. Telephone number 1-800-424-8802.

Odor Threshold - Lowest concentration of substance that has a noticeable odor.

Olfactory - Refers to the sense of smell.

Organic Peroxides - Potentially reactive, often unstable, and explosive organic chemical compounds. (Can form explosive crystals on container surfaces).

OSHA - Occupational Safety and Health Administration, responsible to regulate and enforce safety and health standards in United States workplaces.

Oxidizer - A substance that contains and can yield an abundant amount of oxygen in a fire. These substances can be reactive and increase fire and explosion hazard potential.

Oxygen Deficiency- An atmosphere that contains less than 19.5% oxygen.

Pad (waste management) - A concrete or asphalt surface used for the temporary storage of wastes.

Pathogen - Disease-causing micro-organisms.

Passive Sampling - Atmosphere sampling using natural and not mechanical means to collect samples.

PCBs (Polychlorinated biphenyls) - toxic, "oil like" substances used for many years in electrical transformers and other devices.

PEL - Permissible exposure limit, an established exposure limit established by OSHA regulatory authority. (STEL = short term exposure limit, TLV = threshold limit value).

PH - A chemical method of determining if a substance is acid, alkali, or neutral. Scientifically, pH is a numerical measure of the negative logarithm of hydrogen ion concentration. A measure of pH 7 is neutral, measures less than 7 are acidic, and measures greater than 7 are alkaline.

Penetration - Movement of liquid substances through chemical protective clothing.

Permeation- Movement of vapor or gas through chemical protective clothing.

Plume - Releases of vapor, dust, liquid, or gas to form a cloud of aggregate of hazardous material in the atmosphere.

Plutonium - A manmade silvery "fizzle" metal that is heavier than lead. The half life of Plutonium 239 is 24,000 years.

PMCC - Pensky Martens closed cup flash point test.

Pollution - A contaminate of air, land, water, or articles by substances that are potentially harmful to people or the environment.

Polymerization - a chemical reaction that combines small molecules to form much larger molecules (monomers to polymers). Hazardous polymerization is the reaction that occurs rapidly and releases large amounts of energy - sometimes explosions.

PPE - Personal Protective Equipment, devices worn by workers to protect themselves from hazards in the workplace.

PPB - Parts per billion (1,000 times smaller than ppm).

PPM - Parts per million (1% is equal to 10,000 ppm).

Proper Shipping Information - Per DOT 49 CFR standards, shipping papers should contain: The proper DOT "101 Table" Shipping Name, Hazard Class, UN, or NA #, Packing Group #, RQ and other significant information.

PRP - (A Potentially Responsible Party), persons or organizations that may be associated with environmental pollution and may be required to help pay for cleaning up a waste site.

PSI - Pounds per square inch, at sea level the earth's atmosphere exerts 14.7 psi.

Pulmonary Edema - A reaction in the lungs that causes swelling, filling lungs with fluids, breathing difficulties, and coughing.

Purex - Plutonium uranium extraction, a process used to reprocess spent nuclear fuel and irradiated materials.

Pyrophoric - A substance that is capable of self ignition when exposed to air.

Qualitative Fit Test - Respirator fit test using protocols suggested in the OSHA Respiratory Protection Standard. This test involves checking respirator fit through the use of an odorous product that can be detected by the wearer if the respirator leaks.

Quantitative Fit Test - Respirator fit test using protocols suggested in the OSHA Respiratory Protection Standard. This test involves use of computerized pressure sensitive systems for determining proper fit and fit factor for a respirator installed upon the face of the wearer.

Rad - A unit of absorbed radiation.

Radiation - Transfer of energy through space in the form of particles and waves. Ionizing radiation causes splitting or atomic breakdown. Non-ionizing radiation is from a light source such as ultraviolet radiation which causes external burning and human skin damage.

Radioactivity - Spontaneous emission of radiation from the nucleus of an atom.

Radon - An inert radioactive gas produced by the decay of radium. Radium is linked to the decay chain of uranium 238. Radon occurs naturally in many minerals, a major hazard found at uranium process sites.

RCRA (EPA) - Resource Conservation and Recovery Act - A federal law (1976) enacted to regulate generation, treatment, storage, and disposal of hazardous waste.

Reaction - Chemical change of transformation.

Reactivity - The potential for a substance to transform.

Reducing Agent - A substance that accepts oxygen during a chemical reaction.

Release (Hazmat) - Accidental spillage or otherwise means of pollution generation into the environment.

Reportable Quanitity - An amount of hazardous or otherwise regulated substance that has been released to the environment that must be reported to proper authority. (See references in 49CFR 172.101, EPA list of lists, 40CFR 117.3, 173, and 302.6).

Residue - Dangerous or otherwise regulated substances that remain in an empty container. Depending upon hazardous properties of a substance, some containers are considered empty with as much as 3% of the original substance weight remaining in the container; more hazardous substances must be triple rinsed from the container or inner liners must be removed, see EPA 40CFR and DOT 49CFR references.

Respirator - A device used to protect the wearer from contaminated air. Most common respirators are air purified respirators (APR), self contained breathing apparatus (SCBA) and supplied air respirators.

REM (Roentgen Equivalent Man) - A unit of radiation dosage.

Risk Management Plan - A plan and program established for total planning for and management of hazardous materials incidents.

SAR - Supplied Air Respirator.

SARA - Superfund Reauthorization Act.

SCBA - Self contained breathing apparatus, a respirator that includes a mobile air pack that contains Grade D breathing air. Note: PPSCBA is a positive pressure SCBA, to be worn in IDLH atmospheres.

Secondary Containment - A walled, contained pad or chemical-resistant surface barrier placed around a hazardous substance storage area to prevent hazardous or regulated substance releases to the environment.

Sensitization - An allergic reaction that increases in severity as exposure to an allergen increases. Generally, initial exposures may produce little or no reaction.

Seta - Seta Flash closed cup flash point test method.

Skn - Skin effects (often listed in MSDS).

Solvent - A substance that dissolves another substance; water is the universal solvent.

Solubility - The ability of a substance to dissolve into a solvent.

Solution - A uniform mixture of substances equally dispersed and dissolved in solvent.

Specific Gravity- Weight comparison between a given substance and water. If the volume of a given substance weighs 12 pounds and an equal volume of water weighs 14 pounds, the specific gravity would be: $12/14 = .857$

Stability - The ability of a substance to remain unchanged under reasonably normal conditions.

Stel - Short term exposure limit (ACGIH), maximum concentration allowed for continuous 15 minutes exposure.

STP - Standard Temperature and Pressure.

Support Zone - Also called the "cold" zone or the 3rd perimeter surrounding a hazardous materials work site where activities occur to support work being conducted in the exclusionary (hot) and contamination reduction (warm) zones.

Synergistic Effect - A hazardous effect upon a living system from exposures to combined substances. Combined exposures are greater than the sum effect of each substance alone. In this regard, $1 + 1$ is not equal to 2.

Systemic Toxicity- Adverse health effect upon an entire living system.

Superfund - see CERCLA, Funding for cleaning up abandoned hazardous waste sites.

Target Organ - An organ that is affected by exposures to specific hazardous substances.

TCC - Tag Closed Cup flash point determination test procedure.

TCLO - Lowest published toxic dose.

TCLP (Toxic Characteristic Leaching Procedure) - Determination of hazardous waste characteristics.

Teratogen - A substance that can produce birth defects.

Thermonuclear Weapon - The "H" Bomb, a nuclear device that uses fission to start a fusion reaction.

Thieving Rod - A glass tube used for sampling liquids using vacuum pressure.

Threshold - An established level of hazardous substance exposure. When exposed to amounts below that level, no adverse effects are noted. However, exposures above the established level may exhibit significant adverse health effects.

Title III - (SARA) Community Right to Know and Emergency Response Provisions. Requires users of chemicals to report annual inventories and usage information.

TLV (Threshold Limit Value) - Safe exposure limitations for airborne concentrations of hazardous substances.

TOC - Total organic carbon.

TOC - Also Tag Open Cup, a flash point determination procedure.

Torr - A measure of atmospheric pressure in millimeters of mercury (Hg).

Toxic Substance - A material capable of causing acute or chronic illness or injury.

Toxicity - The sum total of negative health effects from exposure to a hazardous substance.

TPQ - Threshold Planning Quantity (SARA Title III).

Trade Name - A commercial name given to a substance by the manufacturer.

Trade Secret - Confidential information not included on MSDS. Declared by a manufacturer as secret chemical formulation. In the event of an adverse health effect, this information must be provided to a health professional in order that affected employees can be medically treated.

TRI (Toxic Release Inventory) - Under Section 313 of the Emergency Planning and Community Right to Know Act of 1986, annual usage of certain substances must be reported to EPA and other applicable regulatory agencies (Form R Report).

TSCA (Toxic Substances Control Act) - A federal law enacted in 1976 to protect human life and the environment from unreasonable risks associated with manufacturing, distribution, use, and disposal of toxic substances.

TWA (Time Weighted Average) - Hazardous substance exposures over an 8 hour work day and compared to a 40 hour work week.

UEL (Upper Explosion Limit) - A concentration of explosive gas or vapor that will inhibit or "flood out" ignition and sustain burning.

UN Number - United Nations Transportation Hazardous Materials Classification as listed in 49 CFR 172-101. A four digit number that is common to a chemical. Note: if NA prefaces a four digit shipping code, that code is only relevant to North American countries.

Universal Precautions - Assume that everything within a given area is contaminated and protect yourself by using proper standard operating procedures and personal protection.

Vapor Density - A comparison between the molecular weight of a volatile substance and air. For example, the average molecular weight of air is 29 and the average molecular weight of methane is 16: the vapor density of methane = 16/29 or .6.

Vapor Dispersion - Movement of a vapor cloud in air due to turbulence, gravity, spreading, and mixing.

Vapor Pressure - Pressure applied to walls of a container that holds a volatile substance at a given temperature. Temperature and pressure are proportional, the higher the vapor pressure the greater the potential for release into the air environment.

Ventilation - Circulation of fresh air and the removal of contaminated air.

Viscosity - A measure of resistance to internal flow of a given fluid.

Vitrification - Stabilizing nuclear wastes by mixing into molten glass.

Volatility - The property of a substance to mix into air, relative to temperature and vapor pressure.

Vulnerability - Susceptibility of living systems, the environment, and property to potential damage from a hazard or hazardous condition.

Water Reactive - A substance that can produce a hazardous reaction with water.

WMD (Weapons of Mass Destruction) - Biological, chemical, or physical weapons that are capable of destroying the world as we know it.

Weight of Evidence - Evaluation of published information about toxicity and exposure potentials related to a given substance that leads to conclusions about safe handling of the substance. Includes data from available studies, consistency of results and reliability of dose/response testing.

INDEX

Abatement plans, mold and spore-forming
    biohazard contamination, 279 - 282
    Level 1 contamination, 284
    Level 2 contamination, 285 - 286
    Level 3 contamination, 286
    Level 4 contamination, 287
    Level 5 & 6 contamination, 287 - 289
Acid pH, 66
Acids, 66, 169
    health considerations, 176
    safety considerations, 176
Acquired Immune Deficiency Syndrome (AIDS),
    248 - 249 *see also Human immunodeficiency
    virus*
Acts of terrorism and sabotage, 42
Acute exposures to radiation, 90
Aerosol monitoring devices, 100
Aerosols, 62
Air monitoring equipment, 98 - 101, 100
Air moving equipment, 269
Air quality, 193 - 194
Air reactives, 86
Air sampling, 98 - 101, 270
    locations for, 270
Alkali, 67, 169
    health considerations, 176
    metals, 84
    pH, 67
    safety considerations, 176
Alkaline earth metals, 84

Anhydrides, 176
Animal and insect toxins, 81 - 82
Animals, 88
Anthrax, 245 - 246
Asbestos-induced cancer, 212, 214
Asphyxiation, 74 - 75
Asthma, 214
Avian "Bird Virus" genetic variation, 246 - 247
    see also Influenza

Bacterial exposures, 243 - 244
    bacterial spores, 249 - 250 see also Botulism
Bioaerosols, 62
Biocides, 214
Biohazards *see Biological hazards*
Biological agents, 301 - 302
Biological disaster, 301
Biological hazards, 27, 86 - 88, 239 - 240
    abatement plans, 258
    Category A, 302
    Category B, 302
    Category C, 302
    initial evaluations, 259 - 260
    occupational exposure to, 240 - 243
    selected biohazard specifics, 245 - 254
    site inspection checklists, 263 - 264
Blast effects, 322 - 323
Blister agents, 298 - 299
Blood agents, 299
Blood and body fluids, 87

Blood borne pathogens 247 - 248
    standard, 26
Botulism, 249 - 251

Cancer, asbestos-induced, 212
Carbides, 85
Carcinogenesis, 76 - 77
Catalytic ignition, 197
Centers for Disease Control (CDC), 2
Central Intelligence Agency (CIA), 2
Chain of custody, 103 - 104 see also Sample
    records
Chain of evidence, 103 *see also Sample records*
Characterization of hazardous wastes, 9
Chemical, Biological, Radiological, Nuclear,
    Explosive (CBRNE) agents, 294 - 296
    biological characterization, 300 - 302
    chemical characterization, 296
Chemical hazard communication program, 30 -
    32, 30, 33
    in shipping and handling, 34
Chemical Protective Clothing (CPC), 47 - 48
Chemicals, toxic, 214, 218 - 235
Children, protecting, 212, 214 - 217
Chromium (Cr), 211 - 213
Chronic exposures to radiation, 90
Clean Air Act (CAA), 5, 6 - 7
Clean Water Act, 5
Cleaners and disinfectants, 267 - 268
Combustible Gas Indicators (CGIs), 98 - 99
Combustible liquids, 183 - 184, 185
    storage, 185
Communicating hazards information, 22 - 23
Community emergency teams, 129 - 130
    classification and duties, 130 - 132
Community Right to Know, 22, 32 - 34
Comprehensive Emergency Response Plans, 13
Comprehensive Environmental Response
    Compensation and Liability Acts (CERCLA),
    11
Considerations for evaluating safe work places,
    14 - 16
Considerations for safe work practices, 21 - 22
    evaluating safe work places, 14 - 16
Containers and cabinets, 195 - 196
Contingency, 122
Corrosive exposure, monitoring potentials for,
    169

    first aid for, 177
Corrosive hazards, 63 - 65, 161 - 163
    common sense procedures for handling,
    170 - 171
    corrosive air contaminants, 168
    gases, 163
    emergency planning and spill response for,
    173- 175
    health effects from, 168
    liquids, 162
    protection from, 164 - 169
    protective measure checklist, 166 - 167
    safe handling of, 169 - 173
    solids, 162
    storage, 171 - 173
        in bulk, 176 - 177
    use of, 163
Corrosive substances *see Corrosive hazards*
Critical competencies, 73

Damaged containers or leakage, 194
Dangerous substances, 1
    properties of, 61 - 62
Decomposition, 200 - 201
Defective materials, 40
Dehumidifiers, 269
Department of Energy (DOE), 2
Detection and monitoring equipment, 269 - 270
"Dirty bombs" *see Radiological dispersion
    devices*
Disaster, 255 - 256
    and terrorism, 293-295
Disaster site work, 255
    cleaners and disinfectants, 267 - 268
    cleanup - abatement of waterborne biological
    hazards, 258 - 265
    equipment and tools, 268 - 270
    materials & supplies, 267
    putting sites back to normal, 256
    site workers-individual preparation,
    256 - 258
Discipline and training in personnel, lack of, 43
Documentation, value of, 95 - 96
Drying equipment, 269
Dusts, 62, 88, 198

Electrical hazards, 41
Electrical safety, 23 - 24

Electrical wiring in storage room, 193
Electromagnetic Pulse (EMP), 323
Electromagnetic Thermal Radiation (ETR), 323
Electronic charges, 66 - 67
Emergencies, 123 - 125
Emergency action and emergency response
    plans, correlation between, 123 - 124
Emergency personnel, 125 - 128
Emergency Planning and Community Right to
    Know (EPCRA), 12 - 13, 22 - 23
Emergency planning and notification, 12
    and spill response for corrosive substances,
    173
    sample guides, 174 - 175
Emergency responders, 129 *see also Emergency*
    personnel
Emergency Response (ER) plans, 124, 150 - 160
Emergency response Triad of Importance,
    124 - 125
Employee Right to Know, 22
Environmental hazards, 39 - 40
Environmental health concerns, 212, 214 - 218
Equipment operation hazards, 40
Explosives, 319
    and ignitability, 199 - 200
Exposure, 89
    chromium (Cr), 211 - 212
    mercury, 210
    radiation, 89 - 90
Extremes of temperature, 42
Eye and face protection, 46

Fall protection, 41
Federal Bureau of Investigation (FBI), 2
Federal Department of Health and Human
    Services, 2
Federal Emergency Management Agency
    (FEMA), 2, 13
First aid for corrosive exposures, 177
First Responder Awareness Level, 130
First Responder Operations Level, 130 - 131
First Responder Specialist Level, 131 - 132
First Responder Technician Level, 131
Flame-Ionization Detectors (FIDs), 100
Flammable liquids, 183 - 184, 196, 190
    storage, 195
Flash Protective Suits, 50 - 51
Fluid pressure hazards, 42

Food borne diseases, 252 - 253
Food poisoning see Food borne diseases
Fumes, 63, 198
Fumigants, 80
Fungal exposures, 244
Fungicides, 80

Gases, 63
Genus Clostridium *see Botulism*

Hantavirus, 251
Hazard recognition, 34, 36 - 43, 44
    hazard assessment, 45
Hazardous polymerization, 201 - 202
Hazardous Substance Response Fund, 11
Hazardous waste identification, 8
Hazardous Waste Management Resource
    Conservation and Recovery Act (RCRA),
    7 - 10
    management requirement, 10
    permits, 10
Hazardous Waste Operations and Emergency
    Response (HAZWOPER), 121 - 122
    planning, 122 - 125
Hazards Categorization (HAZCAT), 91
Head protection, 45 - 46
Health hazard status, characterization and
    assignment of, 204 - 206
Hearing protection, 47
Herbicides, 79 - 80
High explosives, 200
Human blood and body fluids, 26 - 27
Human health considerations, 72 - 73, 203 - 204,
    206 - 209
    chromium (Cr), 211 - 213
    conditions, 73 - 77
    environmental, 212, 214 - 218
    mercury, 210 - 211
Human immunodeficiency-virus (HIV), 26 *see*
    also Acquired Immune Deficiency Syndrome
HVAC systems, cleaning and decontaminating,
    287
Hydrides, 84 - 85

Ignitable substance hazards, 69 - 71, 185
    authorized storage rooms, 191 - 192
    liquids hazards considerations, 184 - 185
    specifics, 190

storage, 190- 192
 cabinets, 191
 plastic containers, 191
storage room rating and capacity, 192 - 193
safety, 195 - 196
solids, 196 - 197
Ignitability and explosives, 199 - 200
 energies of, 200
Improper lifting and improper ergonomic
 practices, 41
Improper management practices, 43
Improper personal protection, 38
Improper standard operating procedures, 43
Improper work practices, 43
Improperly maintained equipment, 40
Improperly maintained worksite, 40
Incident closure, 128
Incident Commander (IC), 126 - 128, 132
Incident management, 133 - 134
Incident response procedures, 135-147
Indeterminate Public Health Hazard, Category 3,
 206
Influenza, 246
Informing workers about workplace hazards, 17 -
 18
Inorganic chlorides, 85
Insecticides, 78 - 79
International Air Transport Agency (IATA), 2
International Chemical Safety Cards (ICSC),
 185, 186 - 189
International Civil Aviation Organization
 (ICAO), 1, 2
Ionizing radiation hazards, 89
Irritants, 300
Irritation, 74

Job Safety Analysis (JSA), 29 - 30

Lab reports, 272 - 273
Lead, protecting children from, 215
Leakage, 194
Legionnaires' Disease, 251 - 252
Lifting and repetitive motions, 25 - 26
Liquids, 63
 properties of flammable and combustible,
 195
Local Emergency Planning Committees (LEPC),
 12

Lock out and tag out standard, 24
Low explosives, 200

Material forms, 62 - 63
Material Safety Data Sheets (MSDS), 31 - 32
Materials sampling, 101 - 102
Mercury, 210 - 211
Metals, 81
Microbial contamination characterization,
 260 - 261
Microbiological testing, 270 - 271
 interpretation of, 273
Mists, 63, 199
Mold and spore-forming biohazard
 contamination abatement plans, 281 - 282
 sampling, 271 - 273
Monomers, 83
Mutagenesis, 75 - 76

National Aerospace Administration (NASA), 2
National Ambient Air Quality Standards
 (NAAQS), 2
National emission standards for hazardous air
 pollutants, 6
National Fire Protection Agency (NFPA), 2
National Institute for Occupational Safety and
 Health (NIOSH), 2, 14
National Priorities List (NPL), 11
Nerve agents, 298
Neutralization, 67 - 69
Neutron bomb, 321
Nitrides, 85
Noise hazards, 41 - 42
Non compliance of regulatory standards, 42 - 43
Non-ionizing radiation hazards, 89
North American Emergency Response
 Guidebook, 34
Nuclear bombs, 319 - 324
Nuclear fallout, 324

Occupational exposure to biohazards, 240 - 243
Occupational Safety and Health Act (OSH Act),
 14
 OSHA requirements for employers, 14 - 16
 OSHA self evaluation, 19
Off-site survey, audit, and evaluation activities,
 93 - 94

On-site emergency response team leadership, 125 - 127

On-site emergency team training, 129

On-site personnel, 126

On-site survey, audit, and evaluation activities, 94

Organ systems, 77 - 78

Organic peroxides, 83

OSHA enforcement, 16

OSHA Hazard Communications Standard (OSHA HAZCOMM), 16 - 18, 22, 236 - 237

OSHA Hazardous Waste Operations and Emergency Response (OSHA HAZWOPER), 27 - 29

Overhead activities, 41

Overhead and underground utilities, 23

Oxygen and Combustible Gas Indicators (CGIs), 98 - 99

Oxygen, lack of, 39

Pathogenic parasites, 244 - 245

Percent to PPM, 70

Peroxides, 85

Personal health status, 38

Personal Protective Equipment (PPE), 45 - 54, 165 - 166
   classification, 265 - 267
   conditions requiring special personal protective equipment, 48 - 49
   levels of protection, 49 - 51, 50
      OSHA Level A Total Encapsulation Personal Protection, 266
      OSHA Level B Total Coverage PPE, supplied air respirator, 266
      OSHA Level C Total Coverage PPE, air purified respirator, 266
      OSHA Level D Standard Worker Protection Equipment, 265

Personal sampling, 96 - 104

Pesticides, 214

pH, 64

Phosphides, 85

Photo-Ionization Detectors (PIDs), 99

Physical and site hazards, 38

Plague, 252

Plant toxins, 82

Pneumatic hazards, 42

Pollution Prevention Act, 13 - 14

Pollution Prevention Ethic, 13 - 14

Post Closure Liability Fund, 11

Professional, The, 92

Protection from hazardous energy, 24

Public health hazard, Category 2, 205 - 206

Public water management, 7

Pulmonary agents, 299

Pyrophoric substances, 197

Radiation hazards, 88 - 90, 3323

Radiological dispersion devices, 315 - 320

Radon, 215

Reactive hazards, 82 - 86

Recognizing hazards, 29, 30 - 32

Reconnaissance personnel, 94

Record keeping, 10

Regulatory Agency Notification, 10

Regulatory control, 1 - 2

Regulatory programs pertaining to hazardous materials work sites, 20 - 21

Reportable Quantity or RQ, 1, 9, 11

reporting, 10

Respirators, 52 - 53, 54

Respiratory paralysis, 75

Respiratory protection program, 51 - 53
   OSHA respiratory protection standard guidelines, 265 - 266

Risk assessment, 104 - 105

Rodenticides, 80

Routine site categorization activities, 93

Safe Drinking Water Act (SDWA), 7

Safe Work Practices, 19 - 22

Safety, 206 - 209
   ignitable substance, 195 - 196
   safe & prudent work practices, 262 - 263

Sample collection specifics, 103

Sample records, 103 - 104

Sampling plan, 102 - 103

Sampling techniques, 97 - 98
   typical methods conducted for direct examination, 279

Second entry personnel, 95
   selection, 47 - 48

Senior Response Official (SRO) *see Incident Commander (IC)*

Sensitization, 74

Site categorization, 92 - 95

Site configuration, 39
Site inspection checklists, 263 - 264
Site sampling, 96 - 104
Site-Specific Emergency Response Plans, 125
Site Specific Health and Safety Plans (HASP), 29
Smallpox, 252 - 253
Solids, 63
Solvents, 81
    room storage, 193 - 194
Sources of biohazards, 87
Space configuration, 39
Spill Prevention Control and Countermeasure Plans (SPCC), 4
Spills and releases, 27 - 29
Standard Operating Procedures (SOP), 55 - 59
Storage, 190
    containers and cabinets, 195 - 196
    of corrosive substances, 171 - 172
    damaged containers or leakage, 194
    electrical wiring in storage room, 193
    general purpose public warehouses, 194
    in bulk, 172 - 173
    solvent room storage, 193 - 194
    ventilation and air quality, 193
Structural integrity, 40
Superfund Acts, 10 - 12
Superfund Amendment and Reauthorization Act (SARA), 11 - 12
Synergism, 74
Systemic poisoning, 75
Systemic toxicity, 77 - 82

Teratogenicity, 76
Terrorism, 293
Thermonuclear weapons, 321
Title I, 11 - 12
Title II, 12
Tobacco smoke, 215
Too much oxygen, 39
Tools and heavy equipment, 24
Toxic and health hazards, 71 - 72
Toxic atmosphere indicators, 99
Toxic industrial chemicals (TICS), 303 - 307
Toxic substances, 78
tracking, 10
    training, 45
Tularemia, 253

Type A.1 Clean Water, 259
Type A.2 Contaminated Water, 259 - 260
Type B Hazardous Water, 260
Typical pesticides, 78 - 79
Ultraviolet (UV) radiation, 216
U.S. Department of Transportation (DOT), 1
U.S. Environmental Protection Agency (EPA), 1, 4
U.S. Occupational Safety and Health Administration (OSHA), 1
Unstable substances, 82
Urgent Public Health Hazards, Category 1, 205

Vapor density and inadequate ventilation, 71
Vapor pressure, 70 - 71
Vapors, 63, 198 - 199
Ventilation, 164, 193
Viral Hemorrhagic Fevers (VHF), 253
Virus exposures, 243
Volatility and reactivity, 65

Walking and work surfaces, 23
    slips, trips, and falls, 41
Warehouses, general purpose public, 194
Water contamination, 216 - 217
    levels of contamination, 277
    site inspection checklists, 263 - 264
Water damage
    classifications, 259 - 260
        Type A.1 Clean Water Intrusion, 276
        Type A.2 or Type B Contaminated Water Intrusion, 276
    clean up examples, 275 - 276
    evaluators, inspectors, and project managers, 275
    mitigation checklists, 275 - 276
Water pollution control, 4
Water reactives, 84
Water removal equipment, 268
World Health Organization (WHO), 2